DO NOT REMOVE
CARDS FROM POCKET

Engineers and Economic Conversion

Patricia L. MacCorquodale   Martha W. Gilliland
Jeffrey P. Kash   Andrew Jameton
Editors

Engineers and Economic Conversion

From the Military to the Marketplace

Springer-Verlag
New York Berlin Heidelberg London Paris
Tokyo Hong Kong Barcelona Budapest

Patricia L. MacCorquodale
Associate Professor
Department of Sociology
University of Arizona
Tucson, AZ 85721 USA

Martha W. Gilliland
The Graduate College and
    the Department of Hydrology and
    Water Resources
University of Arizona
Administration, Room 322
Tucson, AZ 85721 USA

Jeffrey P. Kash
Graduate Research Assistant
Department of Political Science
University of Arizona
Tucson, AZ 85721 USA

Andrew Jameton
Associate Professor and Section Head
Department of Preventive and Societal
    Medicine
University of Nebraska Medical Center
Conkling Hall, Room 4029
600 S. 42nd Street
Omaha, NE 68198-4350 USA

With one figure.

Library of Congress Cataloging-in-Publication Data
Engineers and economic conversion : from the military to the marketplace
    / [edited by] Patricia L. MacCorquodale . . . [et al.].
        p.   cm.
    Includes bibliographical references.
    ISBN 0-387-94005-7
    1. Engineering — Economic aspects — United States.   2. Economic
conversion — United States.   I. MacCorquodale, Patricia.   II. Title:
Engineers and economic conversion.
TA23.E55   1993
338.4′76233′0973 — dc20   92-40393

Printed on acid-free paper.

Production managed by Hal Henglein; manufacturing supervised by Vincent R. Scelta.
Camera-ready copy prepared using editors' Microsoft Word files.
Printed and bound by Edwards Brothers, Inc., Ann Arbor, MI.
Printed in the United States of America.

9 8 7 6 5 4 3 2 1

ISBN 0-387-94005-7 Springer-Verlag New York Berlin Heidelberg
ISBN 3-540-94005-7 Springer-Verlag Berlin Heidelberg New York

To our children, Robin and Nathan Gilliland, Rachel Jameton, and Katalina MacCorquodale, and the children of the world, in the hope that you may see in your lifetimes what we have not: a peacetime economy.

Preface

The United States' economy is in transition. How much and how rapidly it changes will determine the nation's future well-being. "Economic conversion," that is, a substantial shift of resources—labor, equipment, facilities, skills, and money—from military to civilian production and use is inevitable. How effectively that conversion occurs will be a major determinant of U.S. success in adapting to the new world economy. Although, to date, the rhetoric concerning cuts in military spending far exceeds its reality, large cuts will certainly occur. Unfortunately, where the money from defense will go is unclear. Proposals call for putting the money into debt reduction, environmental protection, new social programs, economic development, and almost every other special-interest area. A self-conscious effort should be made to analyze the nation's needs and establish priorities for where conversion money should go.

This project enjoys the benefit of having focused on the issue of economic conversion before realists considered such reallocation possible. The dramatic changes in the political landscape that have occurred since 1989, when this project was conceived, and today have dazzled us all. Those involved in the project, while holding to the same basic assumptions, have been perceived by others as being, at various times, hopeless idealists and potentially dangerous radicals. The study was conceived at the first board meeting of American Engineers for Social Responsibility (AESR), which occurred in March 1989 in Manassas, Virginia. At the time, Gorbachev, Perestroika, and Glasnost were very much alive. Cold War relations between the U.S. and the U.S.S.R. had begun to thaw, but the Berlin Wall was intact. Anyone predicting substantial reductions in military expenditures was viewed as lacking a sense of reality. As the first drafts of the proposal for support to the National Science Foundation (NSF) emerged, all involved felt NSF reviewers would, at best, view economic conversion as less than serious research and at worst reject it on the basis that we were radicals wasting time studying a future that had no realistic possibility of occurring.

Thanks to the foresight of persons in NSF's Ethics and Values Studies in Science, Technology, and Society Program, and to changes in the international political landscape, the proposal was not rejected. During the period it took to make two revisions of the proposal, the Berlin Wall fell in November 1989, and fledgling democracies emerged in Eastern Europe. The result was that economic conversion moved toward the mainstream. In part due to a funding delay resulting from the federal budget crisis of 1990 (one of many manifestations of the economic challenges faced by the United States), NSF was not able to notify us of support until November 1990. When notified, we wondered whether our ideas, which had seemed so forward-looking just over a year ago, were now passé. But the mainstream character of economic conversion was to be reversed two more times during the project. The Persian Gulf War erupted in January 1991 and hopes for economic conversion collapsed. Following the successful military activities in Kuwait, economic conversion once again moved to the mainstream. Finally, economic conversion was called into question again with the attempted coup in the Soviet Union in August 1991, only to return quickly to center stage with the coup's failure.

Those involved in the study have remained convinced that economic conversion is both a critical need and a potentially great opportunity for the United States to address a pressing set of public policy issues. The need for public policy debate, research, and action on economic conversion has never been more evident. This book focuses on one piece of the conversion, specifically the ability to mobilize and utilize engineering talent and what that means for the engineering profession. The objectives of this study are: (1) to provide knowledge about the impact of economic conversion on engineers, the likely impact of engineers on economic conversion, and the mechanisms that could empower engineers to be prime movers in the economic conversion process; and (2) to identify public policy issues relating to the development and use of engineering talent.

Chapter 1 provides a definition of economic conversion and an overview of trends related to conversion. Chapters 2 through 10 present multidisciplinary contributions to the issues, specifically, how they are seen from the perspectives of economics, psychology, philosophy, engineering, and sociology. In addition, this section offers perspectives from the view of those in academia, the defense industry, and the professional engineering societies. Chapters 11 and 12 synthesize these multidisciplinary perspectives.

Acknowledgments

Special tribute is due Tom Munsey and Jim Evans, who were President and Vice President, respectively, of the American Engineers for Social Responsibility in 1988; initial conversations with them provided the spark to conceptualize this project. The project that resulted in this book was sponsored by the National Science Foundation, Ethics and Values Studies in Science, Technology, and Society Program (Grant No. DIR-9013961). In particular, Rochelle Hollander at NSF was helpful in providing contacts and ideas during the early stages of the project. The ideas and opinions presented in this book are those of the authors and workshop participants and do not represent the positions or policies of the National Science Foundation. Material presented in Chapters 11 and 12 represents a synthesis of ideas generated at a workshop in July 1991. Participants in the workshop included:

Universities
    Lloyd (Jeff) Dumas, University of Texas at Dallas
    Melissa Everett, Harvard Medical School
    Eve Gagné, Chicago State University
    Martha W. Gilliland, University of Arizona
    Andrew Jameton, University of Nebraska Medical Center
    Jeffrey Kash, University of Arizona
    Patricia MacCorquodale, University of Arizona
    Evan Vlachos, Colorado State University
    Marvin Waterstone, University of Arizona
    Joel Yudken, Rutgers University
Defense Industry
    Edward K. Spaulding, Consultant, Hughes Aircraft
    Steven Taylor, Hughes Aircraft
    Gene Vitale, Hughes Aircraft
Council on Competitiveness
    Kent Hughes
U.S. Department of Defense
    Ken Matzkin
Local Government
    Karen J. Heidel, City of Tucson

Professional Societies
American Engineers for Social Responsibility
    Jim Evans, University of Washington (Emeritus)
    Mike Froebe, Boeing Corporation
    Tom Munsey, U.S. Army Corps of Engineers
    Fred Otte, Dames & Moore
    Jud Pitman, Engineering Consultant
    Tim Salo, Program for Appropriate Technologies in Health
Institute of Electrical and Electronics Engineering
    Charles Eldon, Hewlett-Packard
    Vern Johnson, University of Arizona, College of Engineering
Institute of Industrial Engineers
    William Thompson, AiResearch, Tucson Division
National Society of Professional Engineers—National Institute of
Engineering Ethics
    George Pickering, University of Detroit

In addition, James Hayes-Bohanan, Kathy Arnold, and Linda Donatelli
played important roles as facilitators of small group discussions. Special
thanks to Eve Gagné for categorizing and summarizing the lists of ideas that
emerged from the workshop. Charles Eldon, Edella Schlager, Jeff Dumas,
and Marvin Waterstone provided key insights as reviewers of the synthesis.
        Finally, thanks to Dennis Selder for his editorial work in finding a
clear voice among the multidisciplinary perspectives of the chapters, and to
Angela Repp and Patti Boone for deciphering multiple handwritings,
managing a variety of word-processing software, and translating editorial
comments into coherent prose.

Contents

Preface   vii

Acknowledgments   ix

Contributors   xiii

1. Economic Conversion: Why, What, and When   1
Andrew Jameton, Jeffrey P. Kash, Patricia L. MacCorquodale,
and Martha W. Gilliland

**2. The Macroeconomic and Political Structure of
Economic Conversion**   23
Lloyd J. Dumas

**3. Engineers and Economic Conversion: A Psychological
Perspective**   41
Melissa Everett

**4. Economic Conversion in Perspective: The Values and
Ethics of Engineers**   69
Eve E. Gagné

**5. Engineers' Perspectives on Defense Work and
Economic Conversion**   99
Eugene J. Vitale

6. National Technology Priorities and Economic Conversion   113
Kent H. Hughes

**7. The Labor Economics of Conversion: Prospects for
Military-Dependent Engineers and Scientists**   135
Joel Yudken and Ann Markusen

8. The Engineering Profession and Local Economic Adjustment   193
Kenneth Matzkin

**9. The Sociology of the Engineering Profession:
Engineers and Economic Conversion**   199
Evan Vlachos

10. Economic Conversion and Global Justice: The Moral Issues   217
Andrew Jameton

11. Economic Conversion and Engineers: What Do We Know?   237
Patricia L. MacCorquodale, Martha W. Gilliland, and Jeffrey P. Kash

12. Public Policy Directions   249
Martha W. Gilliland, Patricia L. MacCorquodale, and Jeffrey P. Kash

Appendix A: The Participatory Research Process   265

Appendix B: Selected Engineering Societies   271

Contributors

L. Jeffrey Dumas, Ph. D.
Professor
School of Social Science
University of Texas at Dallas
Box 830688
Richardson, TX 75083

Melissa Everett
Public Policy Analyst
Center for Psychological Studies in the
  Nuclear Age
Harvard Medical School
1493 Cambridge Street
Cambridge, MA 02139

Eve Gagné, Ph. D.
Professor
Department of Educational
  Psychology
Chicago State University
Chicago, IL 60628

Martha W. Gilliland, Ph. D.
Dean of the Graduate College and
  Assistant Vice President for
  Research
Professor, Departments of Hydrology
  and Water Resources and Civil
  Engineering
University of Arizona
Administration Building, 322
Tucson, AZ 85721

Kent Hughes
President
Council on Competitiveness
900 17th Street NW
Washington, DC 20006

Andrew Jameton, Ph. D.
Associate Professor and Section Head
Department of Preventive and Societal
  Medicine
University of Nebraska Medical Center
Conkling Hall Room 4029
600 S. 42nd Street
Omaha, NE 68198-4350

Jeffrey P. Kash
Graduate Research Assistant
Department of Political Science
University of Arizona
Social Sciences 315
Tucson, AZ 85721

Patricia L. MacCorquodale, Ph. D.
Associate Professor
Department of Sociology
University of Arizona
Social Sciences 426
Tucson, AZ 85721

Ann Markusen, Ph. D.
Project on Regional and Industrial
  Economics
Rutgers University
Lucy Stone Hall
New Brunswick, NJ 08940

Ken Matzkin
Office of Economic Adjustment, FM&P
400 Army Navy Drive
Suite 200
Arlington, VA 22202-2884

Gene Vitale
Operations Manager
Building 801, Mail Stop G-10
Hughes Aircraft
P.O. Box 11337
Tucson, AZ 85734-1337

Evan Vlachos, Ph. D.
Professor
Department of Sociology
Colorado State University
Fort Collins, CO 80523

Joel Yudken, Ph. D.
Project on Regional and Industrial
    Economics
Rutgers University
Lucy Stone Hall
New Brunswick, NJ 08940

1
Economic Conversion: Why, What, and When

Andrew Jameton, Jeffrey P. Kash,
Patricia L. MacCorquodale, and Martha W. Gilliland

Introduction

The current political potential for converting substantial amounts of military resources to civilian production is one of the great sources of hopefulness for those concerned about the future quality of human life. Many believe that it is possible to turn the vast productive resources now devoted to military threats and actions, which are essentially destructive, to productive uses in transportation, agriculture, construction, electronics, and inventiveness in improving human life. This hope for a shift to civilian production arises from a number of sources: (1) the breakup of the Soviet Union, signaling the end of the Cold War and reducing the danger of bipolar global conflict. (Although far from obviating the risk of nuclear war, this event reduced the probability of a massive nuclear confrontation.); (2) the increasing economic and cultural interconnections among nations throughout the world, leading to philosophies of national security that rest upon economic and cultural strength rather than military strength; and (3) the increasing recognition of the massive needs of populations worldwide for productive and engineering assistance—the need for a decent standard of living; water, road, and shelter infrastructures; and new technologies that are resource efficient, reusable, and recyclable.

This book represents the efforts of a group of people interested in economic conversion to study a few of the complex problems involved in achieving conversion. Our specific focus is on engineers working in defense industries. And more specifically, project participants explore the values, roles, work experiences, and attitudes of engineers with regard to conversion. Engineers are the focus of the book partly because some of us are engineers; we are interested in their social contribution, their lives, and their futures. More importantly, we focus on engineers because we believe that engineers are a crucial resource for creative technological change and so will play a crucial role in conversion. Engineers will also be affected, perhaps more than any other group, by economic conversion.

Although planners can easily discuss converting an industrial plant from military to civilian use, the actual task of doing so is an immense one that should involve engineers at every level. Existing processes, equipment, and materials need to be engineered for new uses. No new technology can be developed and used in production without analysis, skills, planning, and development by engineers. Any new technology, which users may encounter as a physical object or structure or as a product that consumers hold in their hands, requires both social organization and individual inventiveness to be developed. Engineers are thus key personnel in the conversion of technology from military to civilian projects.

In addition to their skills, engineers also have individual goals that make their work meaningful and affect their view of conversion. These values will affect the ability of engineers to cope with career change, to be motivated to modify their skills, and to revise how they see the meaning of their work. On the one hand, defense industry engineers can be expected to be a promising source of creativeness for new civilian technology; on the other hand, they can be expected to have social views and values that make their defense work satisfying, with the result that they may be unable to make a personally successful transition from military to civilian work.

This chapter includes five sections. The first is a brief overview of the audiences to which the book is directed. The second is a full discussion of the definition of economic conversion. Sections three and four, respectively, summarize what is driving economic conversion and factors that suggest that it is or is not under way. The final section provides a guide to the remaining chapters in the book.

Audience for the Book

This book is directed primarily at engineers working in many roles—basic research, manufacture and design, purchasing and sales, education, and management. The editors and project participants want to invite the interest and enthusiasm of engineers working in defense to work toward conversion. We want to stimulate their hopefulness for the benefits of conversion and to think in an informed way about overcoming its problems. We also want to expose engineers to their self-reflections and the observations of others concerning their values and goals. We want engineers to learn more about the larger view of their profession and its social functions. We would like them to take a lead in working with policy planners and the public in finding new opportunities to turn military production toward civilian use.

This book is also directed to policy planners. We would like them to recognize the critical role engineers must play in developing new technology and the consequent importance of engineers to the health of the economy. We thus would like to see planners working more actively with engineers and their organizations in planning conversion. We would not like to see the technological talents and creativity of any engineer go to waste. If engineering skills are not also converted to nonmilitary enterprises, then we can hardly hope for reductions in military expenditures to have a positive economic impact. We thus want planners to pay close attention to the role of engineers in proposed conversion plans.

We also believe that engineers are an important representative group of the people among many groups affected by conversion—secretaries, janitors, salespeople, administrators, soldiers, drivers, accountants, assembly-line workers, etc. We want to emphasize for planners that conversion will cause significant and rapid change in the lives of many people. Conversion is thus a human process; it is not a process that can be fully measured in dollars, new regulations, or corporate planning. Instead, the basic values, optimism, and sense of participation of a variety of people also need careful attention. Thus, not only the goals of conversion, but also the political processes by which planning for conversion takes place, are important to consider.

We are also interested in reaching others interested in engineers and conversion—business administrators and managers, teachers of engineers, and scholars interested in conversion. We all believe conversion will happen, although we disagree on conversion's definitions, scope, and methods of approach. So, this book is intended to aid the process of conversion. Yet, it is not simply an exhortation on duty and opportunity. Instead, this collection represents the efforts of academicians, engineers and business people to share what they know and theorize regarding the current potential for conversion. We believe that we have brought together in this work a valuable range of research and reflection from a variety of disciplines on engineering and conversion.

What is Conversion?

What is "conversion"? Conversion is the substantial shift of resources—labor, equipment, facilities, skills, money—from military to civilian production and use. Those involved in the discussion of the roles of engineers in conversion express various pictures of conversion, and through their wide-ranging images indicate diverse perspectives on economics, public policy, technological development, and defense philosophy. Several basic conceptions of "conversion" are sketched here.

Reducing Defense Expenditures

Some view conversion as simply a reduction in real military expenditures—a reduction in annual Defense Department purchases and services as measured in dollars corrected for inflation, with a concomitant reduction in the number of defense sector jobs, in the amount of investment in defense manufacturing, in the total inventory of military equipment, in the level of standing military services, or in the number or size of companies working in defense areas. We believe that this conception of conversion is a mistake.

Conversion is not just a *reduction* in military expenditures; it requires a *shift* of resources to other productive uses. Thus, those who shape public policy need to consider how resources saved by a real reduction in defense spending can be redirected to civilian uses. At the same time, the process by which a reduction in defense expenditures takes place and the reasons for reduction will all have a major impact on the nature of conversion. For instance, if a reduction in defense expenditures takes place slowly and in small increments, little planning for conversion is necessary. Workers will change jobs and companies will change products slowly and with a limited impact on their communities. However, if conversion is rapid and on a large scale, then the potential impact is great, the dangers are many, and planning is necessary.

If the main impetus for reduced military expenditures is the federal budgetary deficit, a reduction in expenditures may be seen as a cost-cutting measure, instead of new funds to be applied elsewhere. In this instance, no new funds are made available for economic conversion. Even with reductions in military expenditures due to shifts in defense philosophy, one could envision much of the capital and skills now employed in defense manufacture as being unalterable and inflexible because the skills are not readily transferable to use in consumer manufacturing. In that case, if not used for military expenditure, these skills and capital may be of little use.

If a reduction in military expenditure is motivated by a shift in U.S. defense philosophy and by a commitment to build U.S. economic and cultural resources in proportion to military might, then a reduction of expenditures is more likely to be perceived as part of a larger perspective. However, if the reduction in military expenditures is viewed primarily as a product of the need for military efficiency and lower overall costs, the question of directing resources may simply be answered by reinvesting the savings in further military research and development.

Unplanned Conversion

Some envision conversion as having its potentially most positive economic impact when there is only minimal direct public involvement or planning. They believe that the release of military resources will necessarily create new opportunities. Resourceful individual engineers will retrain and seek new positions; flexible companies will retool and create new products for the civilian economy. Those with this position admit that this may take time and that immediate dislocations may have harmful effects for which communities must plan. However, they believe that only minimal temporary assistance should be provided during the time those involved in conversion make changes. But in the long run, no planning is needed.

Should this automatic shift be possible, conversion could take place relatively simply. No planning commissions, no federal programs to create jobs, and no reexamination of the public's needs would be needed. Most importantly, no potentially divisive debates would need to be pursued about the direction of the U.S. economy and its place in the world. Conversion, according to this point of view, could remain a business matter, not an ethical or political concern. To some, this is an advantage of unplanned change; to others, the failure of the public to address these issues would diminish a major source of hope deriving from possible conversion.

In the current sluggish economic times in the U.S., it is hard to feel confident that market forces will provide automatic and unplanned alternatives to defense production. Unemployment among members of the Institute of Electrical and Electronics Engineers is at its highest in twenty years: 5 percent. Reductions in defense expenditures will have to be very slow indeed if they are not to result in depressed local economies and individual hopelessness. Surely, any rapid major reduction of defense manufacturing will require a planned response if the economic consequences are not to be entirely negative.

Toward a Philosophy of Conversion

The nature of this planning—at the levels of individual business goals, economic theory, and public policy—is the main substance of the discussion on conversion. Inevitably, anyone engaged in this discussion must consider the moral and political values that shape the public policy process. Whose interests should most influence conversion planning? What conditions of life should communities seek? What constitutes positive economic development? What sorts of public projects, if this is a desirable use of released resources, would best combine the technological skills of engineers with the needs of the public? What levels of control and planning should be established? To what extent should conversion be a government matter— local, state, or federal—and to what extent the work of market forces, businesses, and professional societies?

The editors of this book believe that U.S. policy planners should take conversion as an important opportunity for an active public policy process, so that the public can take maximal advantage of resources freed by reducing military expenditures. Moreover, the editors believe that a large and rapid reduction of military expenditures can provide an important and rare opportunity for the U.S. to make good use of this large aggregate of resources. Defense manufacture represents an unusually high degree of large-scale, close cooperation between business and government. It is thus possible to envision conversion to large public projects for the public good. Instead of seeing conversion as merely retooling a military electronics factory to produce toasters or home computers, the large-scale nature of military production makes it possible to envision projects to reshape America, such as building new public transit systems, redesigning and rebuilding bridges, building more energy efficient plants, or recycling on a major scale. In envisioning these possibilities, it is important to consider a number of questions, four of which are discussed briefly here:

Whose Welfare?

If the focus of planning is on the welfare of engineers (whom the editors see as especially vulnerable in the process because of their high level of specialization) and other workers in the defense industry, policy would at least direct its concern toward an identifiable group of affected persons. The U.S. could create such policy with readily measurable, short-term results. However, planners then risk thinking in terms of an economic welfare program for workers without considering the public benefits that could be won by addressing the public goods attainable through planned development. Moreover, work simply for one's own welfare is dispiriting work; planners need to encourage conversion to meaningful work.

How Great a Technological Conversion?

How far-reaching should our expectations of changes in military technology be? For instance, consider conversion of nuclear weapons to civilian use. It would take little change in current equipment and skills simply to make nuclear bombs for use in mining or destruction of toxic wastes, as has been proposed by some. Somewhat more major conversion would be needed to put nuclear weapons engineers to work on fusion power production. And an even more far-reaching change to put them to work on solar energy or transportation systems. Conversion with the greatest potential public benefits in the long term may require the greatest changes in companies and engineers.

Whose Organizations?

To what extent should the U.S. strive to maintain the aggregates of companies, relationships, and equipment that now exist in the defense establishment? Suppose that a conversion plan is approved by the U.S. Senate to redirect a portion of defense spending to public transit systems or low-cost housing. Is it important that these contracts be tailored to invite companies like Lockheed or Boeing in the work, or should the U.S. strive to encourage new organizations that might use existing equipment and organize employees more competitively toward these ends? And what would be the economic consequences of such active government involvement in the transportation and housing industries? If the government took tax money now budgeted for defense and switched it to projects for public transit, and General Motors won the contract, Lockheed's plants and employees would not necessarily be saved and some negative consequences to military groups would remain.

What Rate of Conversion?

The rate of military reductions could itself become part of the planning. Choosing a long-term direction and making commitments to undertake reductions at a set rate would make it possible to plan for conversion more effectively. Indeed, a philosophy of conversion should link military reductions and civilian production in an integrated and coherent economic and military philosophy. The rate of reduction in military expenditure could then be seen as depending in part on the ability of the U.S. political system to envision and plan for new directions in civilian production.

Additional tensions warranting discussion in a philosophy of conversion include public versus private management of these resources, local or global objectives, and integrated versus diverse economic activities. Even if policy-makers agree that conversion should be planned and directed toward the public good, much hard work is needed to create a clear economic vision, or set of visions, about the direction toward which defense resources should shift.

Shifting Philosophies of National Security

Planned conversion should not be stimulated simply because the U.S. may suddenly have some cash to spare for new projects. Conversion should be seen as part of a larger picture that includes changes in U.S. philosophy of security linked to a philosophy of civilian production.

What philosophical changes might be envisioned? Presently, U.S. defense philosophy could be said to embody the following five propositions (articulated very crudely here):

(1) A strong military defense is the key component of U.S. security in the world order.
(2) Large-scale production, of which military production is an important element, is essential for the economic welfare of the U.S.
(3) Foreign exchange is necessary to the economic welfare in the U.S. and military weapons sales are a key part of that exchange.

(4) The U.S. must protect militarily its resources and markets abroad, especially energy and oil resources.
(5) A major standing army and weapons inventory are needed for rapid deployment and a converted defense structure would be difficult to mobilize rapidly enough if needed.

A philosophy of conversion would challenge all five of these claims. First, U.S. economic production and exchange can be seen as more important to U.S. security than military might. Second, it has often been argued that nonmilitary production is more productive than military production in creating jobs and maintaining general material welfare. Indeed, the U.S. emphasis on military development and manufacture can be seen as one source of economic stagnation, as is argued by those who point to the economic success of Japan and Germany. Third, the use of weapons as an important element of foreign exchange fosters armed conflict among trading partners and hinders the peace necessary for economic development. Fourth, there is a critical need to reduce the world's dependence on nonrenewable natural resources such as oil. It is thus more important to reduce dependence on oil and other raw materials than to secure availability with force. Fifth, economic strength can be seen as the key to the ability to mobilize forces. If keeping a large standing army and weapons inventory undermines economic strength, then a large military force weakens the ability to respond flexibly militarily and to sustain resistance over an extended period.

These changes in defense philosophy must be linked to changes in economic philosophy. At its most basic level, a commitment to planned economic conversion is a commitment that long-range planning and deliberate public decisions regarding the use of resources can aid economic development. Moreover, a philosophy of conversion must recognize that the world is facing a crisis of overconsumption, exhaustion of nonrenewable resources, waste disposal, and overpopulation. To plan for conversion is to plan toward a redesigned national economy that can produce goods more efficiently, reuse, recycle, and reduce dependence on commodities to fuel material welfare. As environmentalists rightly point out, if the First World does not change its economic ways now, it will have to change them later at much greater cost and with fewer resources.

Many who support work on conversion support this larger shift in public philosophy. Indeed, there is much to be said in favor of a radical shift of economic and security policy and for viewing conversion in this broad way. If freeing of military resources on a large scale happens, then it will necessarily have a large economic impact, and so planners will need to reconsider the role of the military in international economic relations. Nevertheless, not everyone interested in conversion agrees with these views. Some view it as a simple economic opportunity for American businesses to be more competitive in world markets; others view it as a threat to local communities and as a reason to provide for the welfare of workers laid off by reductions in the scale of military manufacture. Thus, the issue of conversion is an important point of debate over American values, economic planning, and visions of the public good for the future.

Conversion of Technological Skills

Whatever one's picture of conversion, technological change must play a central role. And since technologies shape work patterns, profits, and the nature of products, values are inevitably involved in choosing technological directions. For instance, public policy could influence which technologies are focused toward public transportation—more efficient automobiles, more widespread bicycle use, more energy-efficient public transit systems, bigger and faster airplanes, or horizon technologies such as monorails that exploit superconducting materials. Such choices in direction involve policy makers in choosing social and economic priorities.

The U.S. has its best chance of improving the material welfare of the population if it supports public and professional debate on what values might best be employed in converting defense technology to civilian use. If the U.S. fails to engage in serious debate over values in technology because the public believes that values are inevitably diverse, no great or valuable project can be undertaken. Since engineers will have a key role in shaping new projects and products, their values and social vision are important to this debate. Thus, engineers and their personal objectives in approaching new problems must necessarily have a large role in this discussion. The ability of engineers to be able to convert personally from the aims of military production to civilian aims will be critical to the ability of the nation to convert substantial amounts of military production to civilian production.

It is commonly observed that conversion is "a reality" or that it "is happening" or that it "must happen." Military expenditures are decreasing somewhat. Some engineers are now doing different work. Does this mean that conversion is taking place? As the preceding discussion indicates, how observers answer this question depends on the degree to which resources are being shifted to civilian production. However, it is clear that little in the way of a *planned* shift is taking place. The debate over the nature of conversion is thus not just a debate over whether it is taking place or not. It is also very much a debate over what conversion *should* be. The conversion debate thus essentially involves the U.S. public in a debate about its ethical, economic, and political values in relationship to technology and engineering.

What is Driving Economic Conversion?

Economic conversion is being driven by a combination of interlinked political and economic factors.

Political Factors

The threat of global war—and therefore the need for weapons and troops—is widely perceived to have diminished substantially. Events, such as the fall of the Berlin Wall, the Strategic Arms Reduction Treaty (START II, which deactivates a large number of U.S. medium-range nuclear missiles), Soviet democratization, the failed Soviet coup, and the dissolution of the Warsaw Pact and the U.S.S.R. may force a redefinition of international military alliances.

The future role of U.S military forces is the subject of great debate. Military actions, such as the war with Iraq, are seen by some U.S. military planners as anomalies. For example, General Colin Powell was quoted in the *Army Times* as saying "I would be very surprised if another Iraq occurred." In the same interview General Powell stated, "I am running out of demons, I am down to Castro and Kim Il Sung."[1] The local skirmishes that loom on the horizon are less likely to involve a major power now that the Cold War has ended.

Reductions in traditional threats leave U.S military officials searching for new roles for old weapons. Billion dollar weapons development programs, such as the B-2 stealth bomber and the Bloc III tank, were designed in response to Soviet threats in Europe. War between the newly independent republics that comprised the Soviet Union and U.S. forces is unlikely, yet Pentagon officials are still supporting expensive weapons programs.[2] Definitions of the new role of the U.S. military may coincide with large-scale defense spending, but an overall smaller defense budget and scheduled troop reductions indicate smaller U.S. military expenditures in the future.

In addition to political pressures within the federal government, grass roots political pressures call for economic change. A recent editorial in the *Arizona Daily Star*, a Tucson-based newspaper, reflects a common belief that whatever savings come from defense will be used to further political goals of Republicans and Democrats on Capitol Hill instead of addressing long-term problems such as health care or reducing the national debt. Such public skepticism toward government's ability to solve the United States' economic problems illustrates the public pressure for local involvement to revitalize the economy using defense funds.[3]

Economic Factors

Three interrelated economic factors are driving conversion.

Military R & D and Economic Productivity

Widely accepted is the fact that research and development (R&D) plays an important role in product development and economic growth. Lloyd J. Dumas (see this volume) examined the negative and positive economic impacts of military-oriented R&D. He concluded that the use of engineering resources for military-oriented R&D has an overall negative impact on economic productivity. Focusing engineering skills on military technology represents a scientific "braindrain" from the civilian sector. Although positive "spin-offs" into the civilian sector do occur from military R&D, they are marginal compared to the negative effects of the "braindrain."[4]

Dumas notes that the main focus for military R&D is the development of products that will perform in combat situations, while civilian R&D is interested in developing products within cost constraints. Retraining engineers to emphasize cost minimization instead of performance maximization must precede transfers between the military and civilian sectors. Progressive changes in the training and education of engineers may have made this less of a problem than it was ten years ago.

Finally, Dumas argues that military-based industries limit the availability of labor and capital for civilian production. Since cost is not a major factor in military-based industries, they can pay higher wages and prices for personnel and equipment. Civilian companies cannot compete with military expenditures, resulting in the military sector limiting the quality of personnel and equipment available for the production of consumer goods. The three factors of "braindrain," cost-insensitivity, and unequal competition for the available labor and capital all contribute to Dumas' conclusion that the focus on the military sector has a negative economic impact on the production of consumer goods and services.[5]

Federal Debt

Over the last decade the U.S. national debt has tripled in size from approximately $1 trillion to $3.2 trillion . Military spending is a prime target for major cuts in Federal spending. The increasing debt and associated increasing interest create pressure on the federal government to find ways to balance the budget. The underlying assumption is that a balanced budget is expected to encourage economic growth.

U.S. International Trade Competitiveness

U.S. competitiveness in world markets has eroded. The trade surplus that characterized the period from 1900 to 1970 was replaced by a deficit that amounted to approximately $100 billion by 1991.[6] During the 1980s, increases in defense manufacturing jobs accompanied by decreases in nondefense manufacturing employment significantly altered the manufacturing sector of the U.S. The U.S. needs to take steps to revitalize its economy and reestablish the sale of products in foreign markets. Regaining competitiveness in high-tech areas takes engineering talent, R&D, and dedicated resources. Economic conversion could play a role by utilizing resources freed through military cutbacks.

Is Economic Conversion Under Way?

Under the definition, economic conversion is the planned shift of significant resources from military to civilian applications; both the "from" and the "to" components of the shift must have begun for economic conversion to be under way. The "from" half of the conversion is imminent and may have begun. Specifically, the momentum is building toward substantial cuts in the military budget. However, these are early trends and could be reversed relatively easily and quickly. The "to" half of the conversion equation has not begun and discussions are few. There is no public or private mechanism addressing the issue of what we are converting to. No one is responding with a model for how the conversion process should occur nor with a plan for using the resources freed by reductions in the military budget. If there are substantial and significant cuts in military spending, the magnitude of the conversion problem is staggering and planning is a necessity. Initiatives at the federal government, state and local governments, and private sector are summarized.

Initiatives at the Federal Level

Cuts in Defense Spending

In FY 1991 Congress cut the defense budget by 10.5% in real terms although the $286 billion budget was still considerably higher than the Cold War floor of $215 billion (in 1991 dollars).[7] The estimated defense budget authority for FY 1992 was $290.8 billion and $290.9 billion for FY 1993. Although this is not a cut in absolute dollars, this steady state means that the future rate of growth of defense spending could be below the rate of inflation. Planning in August 1992 for the FY 1993 defense budget (the year beginning October 1,

1992) was for a $275 billion level, suggesting that more dramatic reductions are likely. The Department of Defense (DoD) predicts that from FY 1991 to FY 1997 at least $410 billion will be saved and that as many as 2.5 million defense-related jobs will disappear.[8] The majority of savings will be achieved through cuts in the size of military forces and weapons programs.

Scheduled Base Closings

In 1988 Congress selected 86 bases for closure.[9] Bases from this first round of base closure legislation are involved in some form of conversion planning. Examples of such bases include: Pease Air Force Base, New Hampshire; Chanute Air Force Base, Illinois; Mather Air Force Base, California; Norton Air Force Base, California; George Air Force Base, California; the Presidio, California; and Jefferson Proving Ground, Indiana.[10] In addition, the 1990 Base Closure Act led to the formation of the Defense Base Closure and Realignment Commission. The commission identified another 35 domestic bases for closure.[11]

The Office of Economic Adjustment

The governmental response to reductions in military budgets has been to focus assistance at the local community level. The U.S. Department of Defense, for example, established its Office of Economic Adjustment (OEA) in 1961.[12] For thirty years, the OEA has provided financial and planning support to communities affected by base closings or expansions, as well as communities affected by job losses because of cuts in defense contracting. Most recently, the Defense Economic Adjustment, Diversification, Conversion, and Stabilization Act of 1990 designated over $200 million for the OEA to use for economic adjustment and conversion planning.[13]

Federal Legislation

In 1991 numerous pieces of legislation were introduced into the U.S. House of Representatives and the U.S. Senate dealing with the redistribution of defense funds. Such bills include: The Defense Economic Adjustment, Conversion, and Reinvestment Act sponsored by Representative Mary Rose Oaker; The Defense Industrial Diversification and Community Assistance Act introduced by Congresswoman Rosa Delauro; and The Defense Industrial Stabilization and Community Transition Act sponsored by Senator Christopher J. Dodd.

The activity on Capitol Hill did not result in the passing of legislation that could be considered a strong movement towards a comprehensive economic conversion policy. Criticisms of the U.S. House of Representatives 1990 Defense Authorization Act include: the need for more funding; the focus of conversion planning and funds into the hands of the

Department of Defense instead of other federal agencies, such as the Department of Commerce; and the focus on developing dual-use technologies that may have limited commercial applications.[14] However, increased discussion of elements of conversion policy could be interpreted as a first step towards an overall conversion policy. Of the bills that were introduced none has been enacted to date. The following is a brief overview of legislation related to economic conversion.

In 1991, Congressman Ted Weiss, for example, introduced the Defense Economic Adjustment Act, HR 441. It called for the creation of alternative use committees at defense plants, which would develop conversion plans for defense facilities. Other actions included in the bill are the formation of the Defense Economic Adjustment Council, the advance notification of facilities affected by defense cutbacks, and adjustment assistance for communities where conversion is underway. In 1990, proposed legislation included the Defense Economic Adjustment, Diversification, Conversion, and Stabilization Act and the Defense Base Authorization and Realignment Act. In 1991, Senator Tom Harkin proposed an amendment to the Health and Human Services Appropriations Bill for 1992 to reallocate $3.1 billion from unobligated Department of Defense funds to health and education programs (e.g., biomedical research, breast cancer screening, research on mental illness, Chapter 1 educational grants, PELL grants, Ryan White AIDS research, child immunization programs, the Head Start program, the low income home-heating program, and a program to educate immigrants in the Southwest).[15]

Other proposed legislation has focused on industrial policy and technological competitiveness. In 1989, Senator John Glenn supported the Trade and Technology Promotion Act of 1989. This bill proposed the formation of the Advanced Civilian Technology Agency to promote the development of technology in civilian pursuits. Additionally, the bill expanded the membership of the National Security Council to include the Secretary of Commerce, the Secretary of the Treasury, and the United States Trade Representative and established a new permanent position, Assistant to the President for Science and Technology. Senator Glenn's proposal would have reorganized the federal structure responsible for promoting international commerce and technology by, for example, replacing the Department of Commerce with a Department of Industry and Technology and forming a new interagency Export Enhancement Committee.

Senator Joseph Liebermann's bill, the Summary of Defense Realignment and Workers Assistant Act of 1990, would have set aside ten percent of the projected defense cuts in 1991 to pay for the following changes: (1) an industrial diversification study that investigates how defense industries can diversify into nondefense production, (2) the formation of a Presidential Council on Economic Diversification that would coordinate federal agencies assisting in diversification efforts, (3) the establishment of the Defense Industrial Diversification Account to provide funding for businesses that want to diversify into civilian production, (4) the establishment of an Office of Small Business Diversification in the Small Business Administration, (5) the provision of economic adjustment assistance for workers who are laid off from defense industries, (6) the establishment of community economic adjustment planning for affected communities that would identify communities affected by defense cutbacks and make these communities eligible for planning grants, and (7) the formation of an Office of Commercial and Defense Product Integration in the Department of Defense.

One innovative approach to redistributing government defense spending was the National Security, Scholarships, Fellowships, and Grants section of the 1991 Defense Appropriations bill. This legislation reserved $150 million of the intelligence budget to fund semesters abroad, graduate fellowships, and international studies. The purpose of this bill was to make U.S. students more aware of different cultures and languages. One of the bill's main supporters, Senator David Boren, believed the bill was needed because "we can no longer define our national security interest in military terms alone." Boren also stated that "our ignorance of world cultures and world languages represents a threat to our ability to remain a world leader."[16]

Although a number of major bills have been introduced into Congress to begin diversification and conversion, none of these bills have been enacted. There are several obstacles to the planned response to military reductions. First, many bills are introduced by legislators whose districts will be affected by the cuts and yet there is insufficient concern in unaffected districts for the expenditures to be authorized. Second, a national plan for economic conversion would call for a reconceptualization of the relationship between the public and private sectors insofar as public funds would be invested plainly in private enterprises. This raises questions as to which companies will benefit, whether they will then have an unfair advantage, and what this type of government intervention suggests for the "free market." Because of these obstacles to planning on a national level, most of the responses to economic conversion have occurred in particular communities or corporations.

State and Local Planning Efforts

In 1991, over half of all defense spending was in eight states.[17] In many communities that are very dependent on defense industries, an organized effort to deal with the economic impact of cuts in defense spending is occurring. Communities in states such as Arizona, Texas, California, Virginia, and New York are taking part in contingency planning. For example, in Tucson, Arizona, the Greater Tucson Economic Council has appointed a Committee on Defense Industry Diversification, supported by a

$90,000 grant from the Department of Defense, to develop a long-term plan for economic conversion. At the state level, the Arizona Council for Economic Conversion (ACEC), through discussions with local business leaders, city government, and educational institutions, is planning for the loss of defense jobs. ACEC is prepared to assist and consult with interested defense companies in choosing alternative products, but many local companies do not yet see the need for such planning. In July 1992 the Arizona legislature passed a defense restructuring incentive package to assist local firms during this time of military cutbacks. This package, which could save Hughes Aircraft $25 million over five years in state taxes, is described by a Hughes vice president as crucial in selecting a site of expanded operations.[18] The stakes are high; Hughes is competing with McDonnell Douglas for a Navy missile contract worth nearly $300 million.

In 1991, Texas Governor Ann Richards formed the Governor's Task Force on Economic Transition. Its purpose was to develop a statewide plan for Texas defense industries if major defense cutbacks occurred. The task force developed a set of preliminary recommendations on how to coordinate information between state and local groups affected by possible defense cutbacks. The goal of coordination is a plan for short-term financial relief and assistance for defense workers. A plan for long-term changes by the task force is also underway. These recommendations will focus on structural changes in Texas' economy, specifically the selection of civilian technologies to which Texas defense industries can convert. The task force is consulting many of the state's major defense contractors, such as LTV Aerospace and Defense, General Dynamics, Texas Instruments, Bell Helicopter Textron, Rockwell International, and Lockheed. These companies are not part of the task force nor do they endorse its recommendations, but their input may reflect a realization that major defense cuts are a possibility. In addition to the statewide planning, local communities including Ft. Worth, Austin, and Beeville are forming conversion planning groups to try to mitigate the impacts of base closings.[19]

One of the regions of the U.S that may be hardest hit by cuts in defense spending is the Los Angeles area. Los Angeles County is projecting the loss of 420,000 jobs and $84.6 billion by 1995.[20] Two proposals have been suggested to ameliorate the effects of defense cutbacks in the Los Angeles area. First, a local task force is trying to get approximately $50 million from the federal government for a program that helps aerospace companies apply their skills in peacetime. Second, the California Venture Development Fund was established. This private fund of approximately $250 million would try to start up new businesses in the Los Angeles area.

Another prediction for California is that 105,000 aerospace jobs may be lost by 1996.[21] Pentagon spending accounts for eight percent of California's gross state product. In defense-dependent states, such as California, redistributing resources from defense industries to other sectors of the economy is difficult because no single industry can take the place of an industry as large as aerospace. The impacts in "gunbelt" regions, such as California, may illustrate the need for consumer-oriented alternatives to defense manufacturing. The conversion process may be an alternative in many of these cases.

Private Sector Initiatives

Currently, no situation exists in which a major defense contractor is converting a large-scale military production facility to civilian production; however, limited examples of smaller defense contractors marketing products for civilian consumption do exist. For example, Kavlico Corporation, which makes sensors for military aircraft, is now selling its products to the Ford Motor Company. Other companies, such as Sanitation of Compton, who previously provided lavatories for Air Force jets, and Perceptronics, who previously manufactured simulators for the Army, are producing toilets and truck-driving simulators for civilian users like Amtrak and the trucking industry. Many other companies previously involved with the military aerospace industry, including Leach Corporation, are trying to bid for jobs with Boeing when it begins production of its new 777 passenger jet.[22]

One defense company with an aggressive approach to diversification toward civilian production is Hughes Aircraft. Managers at Hughes believe that shifting from defense to civilian markets involves: defining requirements for products internally (unlike defense production where military specifications define products), entering commercial markets early, designing systems that produce payoffs for customers, developing an understanding of civilian markets and customers, and selecting markets in which Hughes has unique sustainable competitive advantages (e.g., markets where Hughes owns patents in important technologies). The technologies that Hughes is developing for commercial use demonstrate that there are useful civilian applications of defense technologies. These technologies include: optical character recognition systems, smart highways, smart cars, electric cars, automated manufacturing systems, satellite network systems, and liquid-crystal light valve projectors.[23]

A Guide to the Chapters

Because economic conversion has such broad implications for society, a multidisciplinary approach to its analysis provides special insights. Our approach was to organize a workshop involving representatives of various academic disciplines, government agencies, and practicing engineers. The workshop format relied on a highly participative process designed to access the collective wisdom of the participants.   The goal was to create a nonjudgmental collaborative environment characterized by teamwork, which would produce a higher level of analysis.   The diversity of the participants ensured a variety of beliefs and ideas.   The workshop format separated idea generation from idea evaluation and used a problem-solving approach to synthesize knowledge and apply it to economic conversion.   Facilitators in small group settings encouraged dialogue, made assumptions clear, and built consensus by identifying common goals and important high-priority outcomes.   We believe this type of highly interactive workshop can serve as a model for the kind of discussion about economic conversion that needs to occur between engineers and policy makers, in engineering schools, at professional meetings, on the shop floor, and in community town meetings.

Economic conversion and its effect on engineers were addressed from the perspectives of economics, sociology, psychology, and philosophy and by engineers from defense industries, civilian industries, academia, and professional engineering societies.   The book includes nine chapters in which specialists apply their specific disciplinary expertise to the issues.   In Chapter 11, the information from the nine chapters and from the workshop is analyzed and synthesized.   The final chapter highlights the policy issues that emerged during the project.   Readers who are interested in a general overview of conversion may wish to skip forward to these broad treatments of the issues and later return to the disciplinary perspectives.

The nine disciplinary chapters begin with a broad framing of economic conversion. In "The Macroeconomic and Political Structure of Economic Conversion," Dumas characterizes military engineering as economically noncontributive and argues that the brain drain (attraction of the best engineers into defense work) is greater than positive spinoffs (useful civilian applications) of military technology. He discusses the transferability of technology and personnel from defense to civilian work and concludes by offering economic conversion as an opportunity to infuse U.S. industry with the technical talent necessary for economic competitiveness.

Chapter 3 turns to the individual level in assessing how engineers will be affected by and react to economic conversion.   Everett reviews the limited research that has focused on the psychology of engineers.   She notes that engineers are politically conservative geographically mobile workers who see themselves as servants of social consensus.   Unplanned conversion is likely to be traumatic for engineers insofar as job loss and unknown risks are involved.   She discusses the role that professional engineering societies can play in easing this transition, in mobilizing support for conversion, and in redefining security in a global perspective.

Gagné, in "Economic Conversion in Perspective: The Values and Ethics of Engineers," notes that values usually reflect circumstances of life and speculates what effect economic conversion will have on engineers who are displaced.   She concludes that this dislocation will be especially difficult for older or more specialized engineers.   She discusses how greater involvement of engineers with end users could motivate engineers to become involved in economic conversion and calls for more broadly based undergraduate curricula to provide future engineers with the personal skills and humanistic orientation.

"Engineers' Perspectives on Defense Work and Economic Conversion" mirrors the research on the psychology of engineers. In this chapter, Vitale provides a defense industry perspective on the differences between defense work and civilian production. His survey of practicing engineers reveals that most of those who work in defense industries have thought about moving to nondefense work. Although engineers believe that salaries and technology are better in defense work, many see advantages to nondefense work, especially in other factors that are related to job satisfaction.

Chapter 6 returns to the macro perspective in "National Technology Priorities and Economic Conversion." Hughes reviews the international competitiveness of U.S. technologies and frames economic conversion as a chance to change national priorities by shifting activities of defense-sector firms into technologies where the U.S. is strong or competitive in international markets. He also offers innovative suggestions as to how to redefine the mission of the 800 national labs by changing laws about technology transfer and altering their organizational structure.

Yudken and Markusen focus on the economics of the U.S. labor force in their chapter, "The Labor Economics of Conversion: Prospects for Military Dependent-Engineers and Scientists." They provide detailed information on the current employment situation of scientists and engineers, which helps resolve the issue of how many engineers will be affected under various conversion scenarios. They demonstrate that the effects of conversion will be concentrated not only within engineering specialties, but in particular industries, firms, geographical regions, and occupations because of the structure of employment in the defense sector. They argue that government-sponsored readjustment (income and relocation assistance, retraining) and facility conversion legislation can mitigate the effects of conversion and ease the transition in these concentrated areas. They remind us of the enormous public investment our society has made in the education, training, and employment of scientists and engineers and call for redirecting these resources to revitalize three national goals: economic stability, health, and the environment. The chapter "The Engineering Profession and Local Economic Adjustment" exemplifies the kinds of assistance that can be provided to regional areas. Matzkin provides an overview of the activities of the Department of Defense's Office of Economic Adjustment; his discussion demonstrates the importance of involving a broad spectrum of community and industry representatives in the conversion planning process and provides examples of successful conversion efforts.

"The Sociology of the Engineering Profession: Engineers and Economic Conversion" provides a societal view of the transformation that is likely to accompany economic conversion. In this chapter, Vlachos discusses the societal changes that led to the development of engineering and the effects that changes in technology have had upon the discipline. As engineering became an established profession, professional societies, codes of ethics, and professional roles emerged. The changing demography of engineering is likely to influence the content of these professional activities. Because we live in a time of complex and rapid social change, the engineering profession, like other professions, is challenged.   Vlachos concludes that changes in the education and training of engineers and the role of professional societies can ease the impact of economic conversion.

Jameton offers a broad philosophical discussion of "Economic Conversion and Global Justice: The Moral Issues." He begins by comparing aggregate, distributive, and comparative concepts of justice and argues that the contemporary global context requires a comparative concept of justice. Jameton applies models of justice (free market, social welfare, and global) to three scenarios of economic conversion: free-market model, compensation model, and global planning model. Next he assesses whether economic conversion increases or decreases justice under each model. He concludes that because justice is one of the most powerful moral concepts, it is important to understand more about engineers' values and concepts of justice and to include justice as one of the criterion used by planners, and community and national leaders in deciding upon the scope and direction of conversion.

Notes

[1]Stephen Budiansky and Bruce B. Auster, "Missions Implausible: The Evaporating Communist Threat has Challenged Many of the Assumptions that Still Guide U.S. Military Strategy," *U.S. News and World Report*, 14 October 1991, p. 28.

[2]Ibid.

[3]"Committee Chosen to Guide Tucson in Shifting to Plowshare Economy," *Arizona Daily Star*, 11 May 1992, p. D6.

[4]Lloyd Jeff Dumas, "The Macroeconomic and Political Structure of Economic Conversion," *Engineers and Economic Conversion: Roles, Values, and Attitudes*, a workshop sponsored by the National Science Foundation, 15-17 July 1991. Graduate College, University of Arizona, Tucson.

[5]Lloyd Jeff Dumas, *The Over-Burdened Economy: Uncovering the Causes of Chronic Unemployment Inflation and National Decline* (Los Angeles: University of California Press, 1986).

[6]Council On Competitiveness, "Competitiveness Index 1991," Council on Competitiveness (Washington, DC: Council on Competitiveness, 1991), p. 6.

[7]Office of Technology Assessment, *After the Cold War: Living with Lower Defense Spending*, Congress of the United States, Office of Technology Assessment, OTA-ITE-524 (Washington, DC: U.S. Government Printing Office, 1992).

[8]Stephen Alexis Cain, "Analysis of the FY 1992-93 Defense Budget Request: With Historical Defense Budget Tables," Defense Budget Project (Washington, DC: Defense Budget Project, 1991), pp. 1-3.

[9]Jim Wake (Ed.), "Basically Confused: Government Gridlock on Next Round of Base Closings" *Base Conversion News*, Center for Economic Conversion, Mountain View, Calif., 1, No.1 (Summer 1990): 4-5.; Jim Wake (Ed.), "Air Base Conversion," *Base Conversion News*, Center for Economic Conversion, Mountain View, Calif., 1, No.4 (Spring 1991): 4-5.

[10]Jim Wake (Ed.), "Around the Country," *Base Conversion News*, Center for Economic Conversion, Mountain View, Calif., 1, No.3 (Winter 1991): 6-7.; Wake, "Air Base Conversion," 4-5.

[11]Jim Wake (Ed.), "Demystifying the 1990 Base Closure Act," *Base Conversion News*, Center for Economic Conversion, Mountain View, Calif., 1, No.3 (Winter 1991): 2-3.

[12]Jim Wake (Ed.), "Pentagon Office Assists Local Communities," *Base Conversion News*, Center for Economic Conversion, Mountain View, Calif., 1, No.1 (Summer 1990): 6.

[13]Jim Wake (Ed.), "Interview," *Base Conversion News*, Center for Economic Conversion, Mountain View, Calif., 1, No.3 (Winter 1991): 1-8.

[14]Gregory Bischack, "A Program for Economic Conversion," draft paper for the National Commission for Economic Conversion and Disarmament (Washington, DC: National Commission for Economic Conversion and Disarmament, 1992), p. 2.

[15]Jeffrey Kash, Conversation with Senator Tom Harkin's staff member Sandy Thomas, 10 September 1991. Graduate College, University of Arizona, Tucson.

[16]Kenneth J. Cooper, "Foreign Studies Expansion Proposed," *The Washington Post*, Legi-Slate article, No. 144917. 1991, Senator David Boren as quoted by Kenneth J. Cooper.

[17]Office of Technology Assessment.

[18]Jeff Herr, "Futures of Tucson," *Arizona Daily Star*, 12 July 1992, pp. A1-A4.

[19]Jeffrey Kash, Discussion with Dr. L. Jeff Dumas, University of Arizona Graduate College, 7 July 1992.

[20]"Task Force Warns L.A. County Could Lose 420,000 Jobs," Associated Press as published in the *Arizona Daily Star*, 18 March 1992, p. A3.

[21]Russ Britt, "Long, Deep Cuts in Defense Budgets Could Mean War for California Economy: Must Seek Other Resources for Permanent Jobs, Growth," *New York Times* as published in the *Arizona Daily Star*, 16 February 1992, p. A13.

[22]Jim Wake, "LA Aerospace Industry in Tailspin," *Positive Alternatives*, Center for Economic Conversion, Mountain View, Calif., 1, No.1 (Fall 1990): 2-3.

[23]Hughes Aircraft Company, "Focus on Hughes, Volume III, Fall 1991," a videotape from Hughes Aircraft Company.

2
The Macroeconomic and Political Structure of Economic Conversion

Lloyd J. Dumas

Abstract

The work of engineers can be an important force driving the long-term growth of productivity that is critical to economic well-being. The large scale and extended use of engineering resources for military-oriented research and development has interfered with this key macroeconomic process while any positive effects of "spin-off" from military research and development has been overwhelmed by the negative effects of this military-sector "brain drain."

Great differences in the practice of engineering between the performance-driven, cost-insensitive world of military research and development and the highly cost-sensitive world of civilian research and development are an important technical barrier to the smooth transfer of military-sector engineers to civilian work.

Retraining and reorienting engineers are critical to overcoming the technical barrier. A decentralized facility-based process of advanced economic conversion planning is key to overcoming the political and economic barriers.

How will the technical community be impacted by conversion? If left to ordinary market forces, the size of the technical community will sharply decline. An extended recovery period of great struggle will ensue, running fifteen to twenty years. However, carefully decentralized planning would allow a far smoother and more successful transition.

How does the technical community impact conversion? In the past, the technical community has lobbied to prevent arms reduction agreements and to support exotic, high-tech weapons systems, generally obstructing movement to conversion. By developing an agenda for arms reduction that facilitates research and development for major productive civilian research and development projects, the technical community could help to speed conversion.

Encouraging the technical community to support conversion. By educating engineers of the social and economic implications of their work, by shifting federal funding priorities toward support of civilian research and development, and by sponsoring conversion conferences and special journal issues through professional societies, engineers can be encouraged to support conversion.

What are the key gaps in knowledge? More accurate data on numbers of employees, the fields and specialties of military-sector engineers, and tracking employment or unemployment after leaving military-oriented work are needed. The ongoing conversion of engineering schools, the development of retraining curricula, and estimates of financial requirements and funding mechanisms must be further studied. Impacts of these workplace changes on engineers on and off the job are also important.

The Macroeconomic Implications of What Engineers Do

The discovery, refinement, and application of technology is widely recognized as one of the most powerful forces in modern society. It is what engineers (and scientists) do. For better or for worse, technological development shapes many dimensions of the world in which we live, including the present performance and future prospects of that part of society we call the economy.

The core of the economy is the social system responsible for providing the material standard of living. Typically the macroeconomic performance is gauged by the rate of growth, the rate of unemployment, and the rate of inflation. Yet given the central purpose of the economy, these measures are only indirect, sometimes misleading, indicators. Better criteria for evaluating the performance of the macroeconomy are the level of material well-being that it generates and the equity of its distribution among people. By developing new technological knowledge and applications, engineers *can* play a critical role in advancing material living standards and thus in furthering the goal of economic activity. Whether they in fact *do* play such a crucial economic role depends on the nature of the activity to which their talents are directed.

Two kinds of goods and services constitute material well-being. Consumer products and services such as cars, televisions, or haircutting directly increase present living standards, while producers of products such as oil field equipment, metalworking machines, and architectural services directly increase the economy's *capacity* to produce, thus raising the *potential* future standard of living. Because the activities required to produce both kinds of products contribute to the main objective of the economy, we can refer to them as economically "contributive."

Most people's earnings take the form of wages and salaries. Therefore, to raise the standard of living that is not only sustained but also well distributed requires that earnings rise faster than the prices of products people buy. If this can be accomplished, the real income or purchasing power rises. Rising wages and salaries mean rising labor costs, and since for most producers, labor is the largest part of their production cost, producers react to the rising cost of labor by increasing their prices. So the price increases wear away the purchasing power of the higher pay, undermining the progress in the broad-based standard of living.

The only viable long-term solution is to find ways of increasing labor productivity, that is, increasing the output that can be produced per worker. Rising labor productivity allows producers to at least partly offset the rising cost of labor and thus to thwart the cost-push pressure to raise prices. Even though employees are paid more, they are producing more. Labor cost per unit of output increases more slowly than wages rise so that prices will not be pushed up enough to harm purchasing power and material living standards.

The standard of living rises over time only if productivity of labor increases. Labor productivity improves by: (1) increasing the skill and motivation of the workforce; (2) increasing the quantity of the capital (tools) available to workers; and (3) improving product and process technology. Engineers have little to do with the first of these approaches, but it is clear enough that engineers are directly and immediately involved in improving product and process technology. And their work also constitutes a major spur to capital investment: the urge to invest in plants and equipment is much stronger when the new machines embody technological advances that make them more efficient, and not just younger, than predecessors. Thus the work of engineers, embodied in new and better machines, is an important force in raising the quantity and quality of capital as well.

The performance of domestic industry vis-à-vis foreign-based competition also depends on the ability of domestic producers to improve the quality of products while producing them efficiently enough to sell them at competitive prices. This too depends on the advance of technology for both product innovation and labor productivity. Overall then, the work of engineers is critical to the long-term progress of industrial efficiency, international competitiveness, and therefore the material standard of living. This, in turn, is central to the economy's role in society. Unquestionably what engineers do impacts macroeconomic performance.

Not all activities that produce goods and services are economically contributive. Other social needs include moral guidance and national security. For example, the services of judges and legislators, and building courts and jails are a few examples of goods and services involved in carrying out the activities of political governance. Such activities are valuable but they serve purposes other than economic activity that further raise the material standard of living. I refer to such activities as economically "noncontributive."

Any type of good or service, whether economically contributive or noncontributive, needs labor, machinery, and other resources. When productive resources are used for economically noncontributive purposes, they are said to be "diverted" from the contributive economy. Probably the single most important category of resource diverting in the world today—and certainly in the U.S.—is the research, development, and production of military goods and services. Military goods and services are clearly noncontributive since they add neither to the present material standard of living (as do consumer products) nor to the economy's capacity to produce (as do producer products). They serve other social purposes.

Engineers and scientists have worked for military activities in large numbers over an extended period of time. In the U.S., roughly 30% of the national pool of technologists work in the military sector. In the former U.S.S.R., even rough estimates of the magnitude of technological diversion are hard to come by. Most likely, the former U.S.S.R. diverted a larger percentage of technological resources than that of the U.S. into the military sector.

Because of the priority attached to military research and development, both the pay and access to state-of-the-art equipment and facilities have generally been much better in the military sector. Accordingly the "best and the brightest" engineers have disproportionately been drawn into military-serving activity. In most fields of endeavor, a particularly talented individual may be especially productive, but will still have only mild impact on the progress of others in their field. In engineering and science, a particularly talented individual can and often does move the whole field forward, changing the starting point for everyone else. So in a sense, although the U.S. may have been diverting 30% of the engineers to military research and development, that figure might represent, in fact, 40% or even 50% of the nation's technological capacity.

Given the critical economic role of engineers, anything that diverts significant numbers of them into economically noncontributive activities hurts macroeconomic performance. Directing engineers into the military sector reduces the rate of progress of contributive technology, interfering with the process of increasing productivity and thus the efficiency of industry. This, in turn, undermines the competitiveness of U.S. producers and the capacity for generating sustained increases in our standard of living.

It is important to note that it is not what diverted engineers are *doing* that causes economic weakness, but rather what they are *not doing*— contributive research and development. This is key to understanding the potential macroeconomic impact of economic conversion. Were noncontributive research and development causing macroeconomic problems, all that would be necessary to correct it would be to stop it. But if the problems are caused by insufficient contributive research and development, throwing engineers out of work will not help. They must instead be removed from the military sector and carefully reconnected to contributive research and development so that everyone benefits.

Some people argue that military research stimulates contributive civilian technology rather than interferes with it, by generating useful applications.   The results of military research and development *d o* sometimes advance civilian technology.   But there is no doubt that drawing engineers away from directly working on contributive research and development tends to have a negative impact on economic progress as a whole. The economically important question is not whether spin-off effects exist. It is clear that they do. The real question is whether the positive "spin-off" effect or the negative "brain drain" (military diversion) effect is greater.

If spin-off were the more powerful effect, nations heavily engaged in military research and development would demonstrate a considerable advantage in developing contributive civilian technology over nations that do comparatively little military research and development.   We would therefore tend to see nations like Great Britain and the Soviet Union at the forefront of civilian technology, while nations such as Japan and Germany would lag far behind. But of course, the opposite is what we observe.

Furthermore, if spin-off were the larger effect, the huge amount of resources poured into military research and development in the U.S. over the last twenty years would have made the 1990s a golden age of American civilian technology. Instead, throughout the 1980s and into the 1990s the rate of civilian technological progress in the U.S. compared to other countries has been seriously retarded. The sorry state of American civilian technological advance has been widely chronicled in such popular journals as *Business Week* with stories bearing titles such as "The Breakdown of U.S. Innovation" (26 February 1976) and "Vanishing Innovation" (3 July 1978). The governing board of the National Science Foundation documented U.S. decline in *Science Indicators, 1974, and Science Indicators, 1978.* Even the 1985 Report of the Commission on Industrial Competitiveness, specially appointed by President Reagan and heavily weighted with executives from military industry, strongly emphasized the problems of American commercial innovation. They certainly were not talking about the weapons business. No amount of spin-off compensates for the effects of the sustained military brain drain.

Those who advocate the spin-off argument typically refer to the examples of radar or early computer technology that originated during the Second World War or in the early postwar years. But in that era, military and civilian technologies were much more similar than today, where the two technologies are marked by a striking divergence. For example, the gap between a World War II vintage bomber and its contemporaries in civilian aviation is minuscule when compared to the gap between a B2 Stealth bomber and a Boeing 757. This ever-widening divergence has sharply reduced the extent to which research and development generate usable spin-off for civilian technology.

Furthermore, military forces demand the capability for high-performance weapons while paying very little attention to cost, which generates technology that inevitably requires considerable additional development to produce affordable civilian applications. In other words, what limited spin-off potential does exist demands a substantial pool of engineers to make technology available for business. A nation diverting a large portion of its technologists to military research and development may simply not have enough manpower left to make even limited spin-off workable. Unsurprisingly, a significant amount of the spin-off from American military technology that does occur, often after lengthy delay, comes out of Japanese industry.

Another factor inhibiting spin-off is the secrecy that typically surrounds military research and development programs, which inhibits the free flow of information and collegial interaction that is so vital to technological advance. An application cannot spin-off from a military technology to which civilian researchers have no access. Military attempts to close off channels of information involving nonmilitary technologies deemed *likely* to have military application further retard the progress of civilian technology.

In conclusion, engineers play a critical role in the performance and prospects of the macroeconomy. To the extent that their talents are directed to the development and application of contributive technology, they are an important force in raising the material standard of living over time. The extended diversion of large numbers of engineers to any form of economically noncontributive activity, including military research and development, produces a net slowing in the rate of contributive civilian technological progress with serious implications for economic performance over the long run. Thus, orchestrating economic conversion and reconnecting military-sector engineers to civilian research and development activity offers the prospect of reinvigorating industrial efficiency and macroeconomic performance to the benefit of us all.

Barriers to Economic Conversion

The problem of converting engineers from military to civilian research and development is complicated by the considerable difference in job settings. The context of military research is one where the ultimate product that emerges is expected to be capable of extreme performance under extraordinarily hostile operating conditions. The belief that even small increments in performance can be decisive in military encounters drives the requirement that weapons and allied systems be designed with maximum achievable capability. The battlefield is expected to expose equipment to extreme conditions, such as vibration, shock, and extreme temperatures, that create additional pressure for maximum performance. On the other hand, the military pays low priority to how much these systems cost. The world of military research and development is thus both performance-driven and relatively cost-insensitive.

Civilian research and development take place in almost the reverse context. Products must of course perform well, but the engineers do not squeeze every ounce of capability achievable out of their design, because small increments of performance are typically not that important, and most civilian products operate in a much less hostile environment. On the other hand, great attention must be paid to cost or products will not sell in the consumer marketplace, where affordability is a major consideration. Thus civilian research and development typically operate in a cost-driven world with much less pressure for extreme performance.

The differences in the context within which military and civilian research and development take place create pervasive differences in the practice of engineering itself. For example, in optimum civilian design, the engineers strike a careful balance between the requirement of adequate performance and the cost implications of the design. Military engineers, on the other hand, make doubly sure that performance requirements are met. They typically do not even know the relationship between the performance requirements and manufacturing cost. Designing civilian products this way results in ones that are far too expensive to be viable.

Engineers themselves are not responsible for the performance and cost objectives within which they must operate, but they must be capable of operating within those boundaries. Doing so requires different skills and a different outlook for the civilian and military areas. The longer engineers have operated in the military world, the more experienced they will be in modes of practice that are inappropriate and unworkable in civilian commercial marketplace. Considerable retraining and reorienting are necessary to convert military engineers and engineering managers to the civilian context. This does not in any way imply that military engineering is somehow less challenging than civilian engineering or that military engineers are less capable than their civilian counterparts. The question is one of difference rather than degree of difficulty.

Apart from technical barriers that must be overcome to effectively convert engineers from military to civilian activity, some serious political obstacles stand in the way. One of the most important is the legacy of what the Pentagon refers to as "political engineering."

"Political engineering" describes the practice of spreading military dollars among as many important congressional districts as possible. Since the congressional district is not a unit of economic activity or technical capability, this practice seeks to maximize the political influence of the military. With projects located in a wide variety of districts, the military congressional representatives support continued high levels of military spending to "bring home the bacon" (or perhaps more accurately, the "pork") in a continuing flow of military dollars and jobs for their constituents. This strategy also provides a convenient lever to discipline unruly members of Congress who oppose key military projects or to reward supporters by shifting military contracts or base activities. "Political engineering" interferes with any attempts to reduce military activity, regardless of how obvious the need for the cutback might be.

During the 1980s, Chairman of the Senate Armed Services Committee, John Tower, sent out a letter to his Congressional colleagues stating his willingness to consider the arguments that an increasing number of them had been making for substantial cuts in military spending. He asked each of them to provide him with a list of all the military contracts and facilities *in their own districts* that they thought were expendable and should be eliminated. Tower reported not having received a single reply.

In addition to the political implications, the geographic pattern of military *spending* also holds the same implications for economic policy. Military industry and bases are spread throughout the country, but at the same time are concentrated in geographic pockets: Boston-Cambridge, Dallas-Ft. Worth, St. Louis, Los Angeles, the San Francisco Bay area, and Seattle are examples of concentrated pockets of military industry; San Diego, San Antonio, Albuquerque, Colorado Springs, and Norfolk, Virginia are locations of military bases and facilities.

In general, stimulating the macroeconomy will help to encourage its expansion and thus create an economic context more conducive to smooth and effective conversion. But conventional macroeconomic policy is not enough, because such a policy would average its effects across the nation. What is required are policies that will reach specifically into the industrially and geographically concentrated pockets of military activity. In addition, general macroeconomic policy will not prepare specialized military capital and labor (including engineers) for absorption into expanding civilian activity, which requires a more structural approach to conversion.

Overcoming Barriers to the Economic Conversion of Engineers

Accepting that retraining and reorienting are critical to effectively convert engineers for the reasons discussed earlier, there still remains the important question of how to implement such a policy. Four steps can adequately describe this process. First, the necessary curricula must be developed. The retraining process shares some common features for all military engineers: most notably learning to evaluate and appreciate the importance of the cost implications of design. But also many specific differences depend on an individual field, along with the precise origin and eventual destination (as well as it can be known) of each engineer involved. That is, a properly designed curriculum must take into account both what the engineer or engineering manager already knows and is used to doing and what they need to know and do in the new work.

Retraining programs without a clear understanding of where the trainee will work do not have encouraging track records. Therefore curricula must be flexible enough to be tailored to each individual case. The design and operation of flexible curricula is not simple, but neither is it beyond the experience of institutions of engineering education. This process must be rooted in reality, and so engineering faculty in direct consultation with engineers and managers of relevant military and civilian enterprises can best carry out retraining. In some cases it may be more desirable for military engineering divisions of firms with civilian engineering divisions to retrain in-house. Boeing is one example of a company that incorporated separate and successful military and civilian divisions that both employ substantial engineering staffs. In such a company, cross-training is possible, though care must be taken to evaluate whether it is optimal. Even so, the number of engineers per production worker of Boeing's military operations is typically much higher than that of their civilian counterparts. Consequently, it seems unlikely that they would retrain all of their former military-oriented engineering staff, because they have no incentive to retrain engineers that they will lay off. Retraining programs at educational institutions will thus have to deal with at least some of the former military-oriented engineering staffs of these companies as well.

The second step involves the institutions of engineering education. Are they presently equal to the task of reorienting engineers? The capacity of these institutions is sensitive to the rate at which converting engineers takes place. Clearly educational institutions would be hard pressed to deal with the deluge of new students that would result from an attempt at sudden and complete conversion of the nation's pool of military-sector engineers. But such a scenario is unlikely. Consider that from half to three-quarters of the pool of engineers may be released from military jobs over the next decade. Estimates of how many of these will retire and how many will move completely out of engineering would be helpful, but the large majority have to be converted or we will lose the enormous social investment that has been made in developing their skills and capabilities. And the engineers themselves will experience considerable personal hardships.

Another problem with engineering schools is that previous emphasis has been on preparing graduates for military engineering. For example, many programs encourage high degrees of specialization, while attention to cost is ignored or de-emphasized. This is appropriate to gearing engineers to operate as part of massive technical staffs in the performance-driven, cost-insensitive environment of the military. However, the degree of specialization and the lack of attention to cost are deadly in the engineering environment of businesses. So institutions of engineering education may themselves have to be converted before they become an effective part of the conversion process.

In addition to considerations of cost, the curricula may be inappropriately structured. Engineering schools, especially the premier institutions, have also acquired faculties whose skills are more attuned to the needs of the military sector. After all, these institutions have been following this path for decades. The faculty may have to be the first to be retrained and reoriented.

The third step of the process is financing the retraining program. Financial aid for education has been substantially cut back during the past decade and is clearly unequal to the task. Present business employers of military engineers do not have a strong incentive to pay for retraining when they can hire inexperienced engineers from school and not bother with the effort and expense of retraining their present staffs. In fact, bearing the full cost of retraining may well put such companies at a serious cost disadvantage relative to other companies without military contracts with which they will be competing after they have converted. And as argued earlier, such companies are likely to lay off engineers to be cost-effective. Companies certainly won't pay to retrain those whom they do not intend to retain.

Local and state governments may be able to offer some retraining assistance, but an increasing number of these are in financial crisis and are cutting back on a variety of key services. The federal government is, of course, also under great financial pressure, but is the only governmental entity that has anything approaching the financial resources needed to address the problem. If the money made available by deep cuts in military spending were apportioned through some reasonable formula among deficit reduction, providing critical conversion financing, and attending to other pressing and too-long-neglected social needs, the financial support needed to address this problem could be found.

The fourth step is integrating retrained engineers into the civilian economy. The market will take care of this eventually, but eventually can be a very long time. Given the pressing need to reinvigorate the competitiveness of American industry and the importance of engineering talent in achieving this, *eventually* may be too long to wait. At the very least, an informational liaison office could connect available engineers with potential employers.

Institutionalizing a mechanism for advanced conversion planning is an effective way of organically connecting converted engineers to contributive civilian activity, and an effective strategy for overcoming the economic and political barriers created by decades of "political engineering". One possible model for achieving this was contained within a Congressional bill called the "Defense Economic Adjustment Act" (HR 441). The bill had two key components. The most important would mandate the establishment of an independently funded labor-management Alternative Use Committee (AUC) at every military contracting facility and military base of over 100 employees. The AUCs would be responsible for working out details, such as selecting products, evaluating markets, retraining, capital modification, and a financing plan for blueprinting the converting facility and its workforce. AUCs would begin this work as soon as they were established, and would periodically update the plans until they are actually triggered by military cutbacks.

The highly decentralized system of AUCs to actually do the advanced conversion planning would allow them to tailor plans for the particular workforce, capital, and location of the facility by those who know it best. This would overcome the structural economic problem posed by the geographic and industrial pattern of the military sector by reaching directly into the pockets of military-sector activity. It would also help overcome the political barrier that is the legacy of the Pentagon's practice of "political engineering," by reassuring those who work in the military sector that carefully designed, specific plans for retraining, reorienting, and reconnecting them with alternative civilian activity exist when they are no longer needed for military projects.

The second part of the bill would fund income support, maintenance of health insurance, pension plans, and retraining and relocation if necessary for employees of contractors or bases whose military work was terminated. The support would be transitional only and would be available for a maximum of two years. The bill also allowed contractors to set aside special tax-sheltered funds for financing capital modifications required when the conversion plan actually would be implemented.

Making this whole advanced conversion planning process work requires the direct participation of individual engineers. Their input into the search for new areas of civilian work that fit the skills of the workforce and the capabilities of the facility is also invaluable. But professional engineering societies also have an important role to play. They can do a great deal to encourage conversion. Professional engineering societies could establish task forces to lay out a broad agenda for the most promising civilian technological areas within their own disciplines. This would stimulate political debate on the shape of a national research and development agenda and the thinking of the engineers who will be directly involved in conversion planning. Engineering societies could also sponsor conferences and special issues of their journals focused on the conversion of engineers. This should be of great help in legitimizing this activity, in educating their members, and in motivating members to get involved.

The Impact of Conversion on the Technical Community: Two Scenarios

The enormous fiscal pressure on the federal government combined with the disintegration of the Warsaw Pact as a primary military threat are likely to lead to major cutbacks in military spending in general, as well as military research and development funding. Furthermore, any serious attempt to infuse the magnitude and quality of technical resources into the civilian commercial economy needed to rebuild American industrial competitiveness will increase the pressure to release engineers from the military sector. It is therefore likely that the portion of national technical resources taken by the military will drop precipitously over the next ten to fifteen years. What will this do to the technical community in the U.S.? Consider two scenarios.

In the first scenario, no advanced planning occurs for economic conversion. Laid off in large numbers after cuts in military spending, engineers are left to fend for themselves. What would happen then? In the first place, those who can will try to find jobs within what remains of the military sector. This will create an oversupply of labor that will tend to cause a substantial drop in pay for those engineers who continue to do military work. But the vast majority will have to look for work elsewhere. They will naturally try to interest high-technology civilian companies in hiring them, but will discover, much to their dismay, that civilian companies are simply not willing to employ engineers whose main experience has been in a highly cost-insensitive environment. Those civilian companies who do hire them will soon discover that their past experience and expertise are liabilities. Some may undertake the effort and expense of retraining and reorienting them; most will not. They will fire them instead, and exercise greater care in avoiding engineers with such professional backgrounds in the future. A few companies will not recognize the problem and will suffer damage as a result.

Having learned a lesson in the job market, some discharged military engineers will attempt to get themselves retrained to overcome the obstacles they have encountered. They will consult with counselors in engineering schools, try to put together a reasonable course of study on an ad hoc basis, and finance this as best they can. Some will find the right schools, get good advice, have the necessary financial resources, and succeed after an extended period of struggle. Many more will get discouraged, leave engineering, and try to find another job to hold themselves together financially, albeit at a substantially reduced salary. A small number will try to start their own business, but if they enter civilian fields in which they are applying their own unmodified engineering skills, they are likely to experience a high rate of failure. A few will succeed in their entrepreneurial efforts.

Applications to schools of engineering will drop as students react to constant media coverage describing the overabundance of engineers and the personal struggles of former military engineers unable to find work or adequate pay. Just when the nation's need for technical talent to rebuild competitiveness is growing, is when the flow of new blood into these fields will be reduced. American industry will fall farther behind in the competitive global market. Consequently, the standard of living will continue the decline of the past fifteen years, perhaps at an accelerated pace.

Engineering instruction will slowly adjust, and the job market for engineers and the public image of the engineering profession, after some fifteen to twenty years, perhaps longer, will once more run smoothly, a generation of highly talented and skilled engineers having been sacrificed in the process. On the whole, not a very encouraging picture.

In the second scenario, a bill like the Defense Economic Adjustment Act passes. Large military research and development cutbacks occur only after Alternative Use Committees, in which engineering staffs have been prominent participants, develop conversion plans. Once fired or laid off through lack of work, the blueprints for converting the facilities and their workforces are put into action. Terminated employees begin to receive transitional income support, key fringe benefits, and allowances, which they use to pay for retraining programs in which they now enroll. These have for the most part been designed and set up in cooperation with local engineering schools.

Physical modifications in the facilities begin in accordance with the plans while managers and engineering staff are being retrained. The communities in which they live do not suffer the multiplied impacts of the loss in military-sector payrolls because the income of the workforce is being maintained.

A year or two after implementing conversion plans, the workforces are retrained and ready to go, the facilities have undergone necessary modification, and the new civilian activities begin. Some of the conversion plans are successful and need little modification; they are efficient, relative to competitive civilian operations, and can go into high gear. Most of the plans will need to be modified as they are implemented due to unforeseen circumstances and problems that are likely to arise. With some false starts, the problems are corrected and the plans succeed. A few of the conversion plans fail for a wide variety of reasons, ranging from incompetence to unpredictable changes in the economic environment.

The engineers (and other employees) who lose their jobs as the result of failed conversion plans join the ranks of those who could not be fit into the plan in the first place. As mentioned earlier, some fraction of the engineering staff will not be in the newly established civilian activity. Job bank services will be made available to all those who are not absorbed by successful conversions, to help them locate opportunities with other civilian employers. In addition, it is very important to note that the engineers who find themselves in this position will have undergone retraining programs (paid for under the legislation) that have rendered them perfectly competent to operate in the cost-sensitive world of civilian design. Civilian employers will therefore not consider them unusable. Even the engineers who remain in military work gain by avoiding the depressed wages that competition by laid-off military-sector engineers would cause.

The influx of engineering talent revitalizes American industry, and within a decade we begin to win back many of the markets that have been lost to foreign competition. American industry once more becomes a competitive force to be reckoned with, and the decline of average living standards is reversed.

The Impact of the Technical Community on Economic Conversion

The activities of the technical community are important to the process of economic conversion. In the past, military engineers and scientists lobbied to undermine progress in arms reduction, and thus undermined conversion. For example, technical staff from the nuclear weapons labs have sought to convince Congress to continue testing nuclear weapons, thereby thwarting agreement on a comprehensive test ban treaty. More broadly, they have tried to promote military technological solutions for what are really socioeconomic and political problems, from using sophisticated military technology to root out and destroy illegal drug trafficking, to developing systems to defend against missile attack and render nuclear weapons "impotent and obsolete." Neither the drug problem nor the problem of the nuclear arms race are amenable to technical solutions because they are not technical problems. They must be solved in the socioeconomic and political arena.

Promoting technological solutions to catch the public imagination has helped to keep military research and development money flowing, even when the preponderance of opinion among the technical community as a whole was that the solution was technically unachievable (as in the case of the Strategic Defense Initiative). The willingness of some engineers and scientists to accept public money to carry out research under the umbrella of a military program that they personally do not believe is technologically possible raises serious ethical questions.

Further, all the competition for military research and development funding has prevented the nation from turning its attention to other areas of technological work outside the military realm. Ironically, some of these neglected areas of civilian research would demonstrably contribute even more to the nation's security than research and development on weapons systems. For example, converting the national nuclear weapons labs into centers to study ecologically benign energy sources and conservation would do far more to advance our security (by getting us off the international oil pipeline) than would developing yet another generation of nuclear weaponry.

The technical community needs to stop playing this obstructionist role. There is much that engineers can do to speed the rate of conversion. For one, they can design an agenda for research and development to facilitate, rather than undermine, progress in arms reduction. The most obvious area of work along these lines is the development and refinement of improved verification and inspection technologies. The technical community could also develop an agenda for civilian research and development to address the technological component of key areas of economic and social need. Improved technologies for cleaning up the ecological mess, for dramatically improving the quality and efficiency of key industrial machinery and equipment, for reducing the cost of revolutionary medical equipment such as CAT scans and artificial organs so that they are more accessible—these are but a few of many possibilities. Such an agenda, realistically designed and enthusiastically promoted, could surely capture public attention and support at least as effectively as the endless stream of military projects that have been generated in the past.

How can the technical community be encouraged to support conversion? First and most importantly, we must educate engineers of the enormous impact the application of their knowledge has on the economy and on broader society. Most engineers are educated within a fairly narrow technical curriculum. It is often argued that engineers have so much to learn to be effective contributors to their field that there is little or no room for courses that do not directly further their disciplinary training. Yet, as the bearers of immensely powerful knowledge, it is the obligation of engineers to consider the social and economic implications of what they do or fail to do. And they can scarcely be expected to consider these implications of their work if they do not understand larger economical and political issues.

Second, we must work to shift federal research and development priorities toward much greater support for civilian engineering research. As long as the largest single funder of research and development focuses such a large portion of its money on military research, it will be very difficult to generate the excitement necessary to redirect the attention of the engineering professions. Throwing their considerable prestige behind such a shift, the engineering societies could make a real contribution.

Third, many have argued that the professional satisfaction of cutting-edge military research with low cost constraints cannot be duplicated by civilian research. To this I have two responses. On the one hand, it is a greater intellectual challenge, a greater test of professional skill, to successfully carry forward the frontiers of knowledge in a more constrained environment. A researcher must be more creative and more ingenious to get the job done when cost is a consideration. Yes, it can be frustrating, but success is that much sweeter when the test is more severe. On the other hand, it is not just the work itself, but also the nature of the output that makes a job more fulfilling. We tend to underestimate the impact of the nature of output on work satisfaction. For many engineers—if not for all—working on products whose primary function is to destroy people and property may not be as satisfying as doing research and development directed toward more life-enhancing goals.

Gaps in Knowledge, Areas for Further Research

Some data, basic to engineering conversion analysis, are not known with sufficient accuracy. Perhaps the most fundamental is the number of engineers that are currently engaged in military research and development. Various estimates exist, but these estimates vary depending upon the political orientation of the sources, and the range of estimates is far too great. Not enough is known about the economic and professional fate of engineers who have lost their jobs in the military. How many of them are unemployed? How many are employed in jobs that do not make use of their considerable skills? How does the current employment of engineers in military research and development and the re-employment experience of those who have been laid off break down by field and subfield of engineering? By economic sector?

Apart from this, a number of key retraining issues require further work. A basic structure must be developed for a flexible curriculum to retrain military engineers for alternative civilian activity. The issue raised earlier about the nature and extent of conversion of institutions of engineering education that might be required needs further exploration and analysis. Just what changes in faculty, curricula, and perhaps organizational structure will most effectively facilitate the critical retraining? Estimates of the financial requirement for carrying out this retraining must be made, and a variety of financing mechanisms need to be investigated.

Finally, a number of workplace issues require further analysis. Among the most important are the extent to which efficient conversion requires changes in the structure of organizations in which engineers work and the implications of these changes for their work environment. Also of considerable significance are the psychological impact of conversion on engineers and its implications for both work performance and the quality of their lives outside the job.

Conversion as an Opportunity Rather than a Problem

It is all too easy to become caught up in the complexity and difficulty of the conversion process and to begin to see it only as an enormous problem. It is complex and difficult, but it is also a tremendous opportunity to redirect the efforts of some of the most productive resources of society toward solving some of the most critical issues of our day. We should approach the prospect of large-scale conversion of our engineering resources with the excitement and enthusiasm that such an opportunity deserves. But in our excitement, we must not get so wrapped up in what engineers do now that we forget to give sufficient attention to what they can later accomplish.

Many of the most pressing problems of our time have a technological component: environmental protection, energy, improved quality and accessibility of health care, economic development and global security. But none of these are purely technical problems, and developing grand technological schemes to fix them without attending to their key socioeconomic and political dimensions is futile and even dangerous. In the 1950s, great enthusiasm for the complex of agricultural technologies broadly known as the "green revolution" led some to argue that the systematic application of these technologies would finally eliminate hunger from this planet. The green revolution technologies worked—they vastly increased food production. But more chronically hungry people live on Earth today than before the green revolution took place. That is not because increased food production was not important, but because hunger was not solely a technical problem of production but a socioeconomic and political problem of distribution.

Conversion of military-sector engineers is the only source, now available in less than a generation, of technical talent necessary to put American industry back on track and provide a long-term viable base for continuing improvement in the material standard of living. Such a step, although necessary, is not a panacea for our industrial problems or even for our technological needs. Other crucial issues include abandoning the short-termism so common among American managers today and encouraging women and minorities to become full participants in the national community of engineers and scientists.

From a purely technical point of view, the progress of military technology over the past half century of Cold War is nothing short of astonishing: a clear illustration of the talent embedded in the American technical community. Now a new challenge lies before us. Can we redirect this enormous mental capacity from the innovation of ever more destructive weaponry to innovation for an ever more productive economy? Other than a lack of will, it is hard to imagine what could stop us.

3
Engineers and Economic Conversion: A Psychological Perspective

Melissa Everett

Abstract

Impact on the technical community. Depending on the scenario, economic conversion will require modest to substantial adjustments from the technical community. The impact of planned economic conversion would range from mildly difficult to beneficial, depending on how the process is managed and motivated. Unplanned conversion, however, is likely to be traumatic to many engineers and technical organizations, involving job loss and unknown risks. Either way, this impact depends greatly on the degree to which engineers are prepared and involved in choosing alternative futures and the extent to which they can see the usefulness of their talents in a converted enterprise.

Impact of the technical community on conversion. Most leadership and policymaking for conversion will be exercised by others, including many managers who come from the engineering ranks. But engineers are likely to be a "swing vote" in deciding whether, and how coherently, economic conversion will take place. Engineers' impact on a process of planned conversion could range from cooperation and enthusiasm to apathy or active resistance. In any plausible scenario, the extremes of large-scale, dynamic advocacy and open rebellion are unlikely.

Encouraging constructive participation. Engineers can be encouraged to participate constructively in economic conversion by applying the evolving principles of the "psychology of empowerment." According to one model, socially responsible behavior hinges on four elements: knowledge, emotional experience, personal connection to the issue, and patterns of action that reinforce conscious commitment.

Barriers to constructive participation include engineers' political and social ideology, professional responsibility constructed as detachment and obedience, fascination with technology, optimism in predicting technological impacts, personality tendencies (e.g. need for order), and how workplace experience shaped self-image and expectations.

Research agenda. Existing knowledge about engineers' psychology is extremely inadequate. Much is based on anecdotal material, outdated findings, and often small samples. Matters are further complicated by the multiplicity of psychological research models and inconsistencies among the theoretical frames of the studies.

A broad effort by social scientists to study this population—or, better, a movement toward greater self-knowledge on the part of engineers themselves—could alleviate one of this nation's most acrimonious debates. One starting point would be a detailed portrait of engineers' attitudes about military work, personal responsibility, and economic conversion. Beyond this some areas for investigation are:

1. What knowledge and beliefs about the world underlie attitudes toward military work, personal responsibility, and the prospect of economic conversion?
2. How do these attitudes vary with job experience, industry, point in life cycle, political climate, and institutional climate?
3. To what extent do engineers' attitudes about national security and their constructs of professional responsibility stem from innate personality characteristics, and to what extent are they artifacts of work experience and group influence?
4. To what extent does the psychology of defense intellectuals and military-industrial cultures, such as fascination with weapons and minimization of their human impact, apply to engineers?
5. What are some civilian technologies that can fascinate engineers and be useful to society?

Introduction

To begin, here is a qualitative description of the engineering population, written by one of their number who has made a career out of understanding engineers, defending them against critics and telling them to shape up when he considers it necessary. In *The Civilized Engineer,* Samuel Florman describes "the engineering view" as:

> A commitment to science and to the values that science demands—independence and originality, dissent and freedom and tolerance; a comfortable familiarity with the forces that prevail in the physical universe; a belief in hard work, not for its own sake, but in the quest for knowledge and understanding and in the pursuit of excellence; a willingness to forego perfection, recognizing that we have to get real and useful products 'out the door'; a willingness

to accept responsibility and risk failure; a resolve to be dependable; a commitment to social order, along with a strong affinity for democracy; a seriousness that we hope will not become glumness; a passion for creativity, a compulsion to tinker, and a zest for change.[1]

Samuel Florman is an engineer.

This passage reflects some of the limits to what we know about engineers. First, this list of qualities is by no means complete. Florman selects properties emphasizing the cognitive and the impersonal. For example, he describes engineers' relationships to the mechanical universe much more than their relation to people. Secondly, this list is by no means internally consistent. "Dissent and freedom and tolerance" and "commitment to the social order"? "A resolve to be dependable" and "a zest for change"? These may not be full-fledged contradictions, and certainly many of us balance these tendencies. But they reflect tensions and complexities that are among the unknowns before us.

Psychology lacks a Florman to say who psychologists are. Psychology is a fragmented field, with dozens of subcultures, levels of analysis, and theories. Even empirical studies reflect a wide variety of theoretical frameworks, often unstated. From this, the known universe of "the psychology of engineers" can be said to include personality characteristics and self-concepts, political and social beliefs, motives for working, and common experiences on the job. I attempt to synthesize useful elements drawn from psychoanalysis, humanistic psychology, psychology of the self, group dynamics and family systems theory, linguistic analysis, and mass psychology through discussion of some specific studies of engineering populations and some broader social/cultural studies that should be explored for their applicability to this profession.

Finally, I will relate these descriptions to the requirements for engineers in terms of leadership and constructive participation in some likely conversion scenarios.

The relevant psychological questions depend on the scenarios for economic conversion under consideration. The most likely scenarios will require leadership from some and active cooperation from many, both in the technical community and the general population.

Personality Characteristics

The studies considered here suggest that many of the personality traits and work styles of engineers are well suited to contexts where fundamental assumptions are relatively stable and the major challenges of work are technical. Some of these strengths become handicaps in times of shifting social paradigms, when there is enormous interplay between technical questions and questions of value.

For example, according to Haworth, Povey, and Clift, engineers show greater need for order, self-endurance, and achievement than the overall population.[2] Moreover, in a discussion of barriers to interdisciplinary education, Henry Bauer makes the case that the work of scientists and engineers presumes a particular cosmology: "Scientists and engineers believe implicitly in certain absolute truths, and further believe that given enough time and effort the absolute truth can be found."[3]

Some of these characteristics may be related to the life circumstances of many engineers and the history of the profession. Typically coming from backgrounds of less socioeconomic advantage than other professionals, engineers fought for enhanced status a generation ago.[4] Today, members of the profession see themselves as practitioners who have dignity, status, and importance. They often align themselves with the interests of their employers, giving low priority to personal autonomy.[5] In fact, Zussman reports that when groups of engineers in two companies were asked whether they would like to work in a more democratic environment, large majorities said no (70% in one company and 77% in the other).[6]

In at least one engineering population—those at Lawrence Livermore National Laboratory—distinctive characteristics show up in information-processing and decision-making styles. Linda M. Donald administered the Myers-Briggs Type Indicator to a group of Livermore engineers.[7] Donald's subjects stood out strongly on two of the four dimensions measured by the test. First, their decision-making style tended toward "thinking" rather than "feeling," which in the sometimes arcane language of the Myers-Briggs, means a detached approach, grounded in laws of causality and deductive logic, with personal sensibilities and values held at a distance. Secondly, they made sense of events by "judging" rather than "perception." This mode emphasizes planning, control, and predictability, sometimes at the expense of intuitive openness to events as they unfold.[8]

Motives for Work and Sources of Job Satisfaction

Engineers give mixed messages about their reasons for choosing and staying in the profession and the sources of their career satisfaction. One of the most-debated themes is an ostensible shift from idealistic goals in the 1960s and 1970s to narrow and immediate self-interest in the 1980s. A 1968 study found people choosing engineering as a source of "interesting work," whereas by 1986, the primary reason given was "job opportunities."[9] Altruistic reasons such as "helping others in difficulty" also declined in the mid-1980s.[10]

Discussions of this trend often lead to the exaggerated conclusion that engineers are losing their ideals and have forgotten all motivations but greed.   But an in-depth exploration of career anchors by Griggs and Manning concludes that—regardless of engineers' initial motivations for choosing the profession and particular jobs—what satisfies them on the job is "complex technical problem solving and the opportunity to work with stimulating colleagues and to make a meaningful contribution to society."[11] A survey of several bodies of literature on engineers' work values is provided by Watson and Meiksins, who conclude from their own 1986 survey research that "engineers have become highly focused—perhaps overly focused—on the gratifications derived fromtechnical work as a process."[12]

Furthermore, evidence in related professions such as physics suggests that in the job search the pendulum may be swinging away from immediate and material considerations and toward greater emphasis on social contribution.   Reporting in *Physics Today* on 1986 survey responses of undergraduate physics majors, Susanne D. Ellis notes:

> It used to be the case that graduating bachelors hardly ever wrote anything in the part of the questionnaire reserved for comments. But in the last two years, respondents mention increasingly frequently that they are unhappy with the job prospects outside the defense sector. A typical comment was, "The reason I have such trouble finding a job is that I do not want to do defense related work."[13]

This observation fits well with broader trends.   Professionals in general and baby boomers in particular are showing restlessness amid their successes.   *Real Work*, a book in progress by Robert Aaron and Michael Bernick, examines the job frustrations of professionals who are successful but feel that their work lacks meaning. Aaron sees this problem as inevitable in a service and information economy, where even highly skilled work is often compartmentalized and serves an indirect or intangible purpose.   On the other hand, Douglas LaBier, a psychologist who has intensively studied the "urban career culture," believes a change is beginning and that "there is a deep yearning for a new vision of adult life, but there are very few role models for a better way.   People are really redefining the meaning of success."[14]

The extent to which this redefinition will lead professionals into or out of engineering is unknown.   One factor is the degree to which the profession is seen as a true servant of social needs rather than as a servant to the interests of solely industry or science.

Experience of Work

Personality and values are not static, but are shaped by life experience. Insight into engineers' experience on the job comes from a mix of academic research and the popular press.

Engineers are mobile in pursuit of their careers, averaging less than four years with a company.[15] Their perception of the extrinsic rewards of technical work become more negative with experience.[16] Many engineers feel that their skills are underused by bureaucratic managers who assign them narrow and repetitive tasks.[17] These findings are all the more significant because engineers' reality-based expectancies shape their values, especially in organizations that have a reward structure.[18] Understanding the relation between expectations and experience is important to assess the degree to which engineers' professed beliefs—about conversion or anything else—are immutable or subject to influence.

In the last few years, media coverage of the challenges facing the engineering work force has been discouraging: there are not enough engineers, and their companies and industries are not competitive.[19] Engineers hear that they are not adequately trained in cutting-edge technologies and that they often fail to rise to the basic challenges of professional responsibility.[20] The effects of such messages on engineers' self-esteem, group loyalty, job satisfaction, and perception of their choices are important areas for examination.

The stresses on engineers may well be compounded by a commitment to objectivity, which emphasizes keeping one's personal reactions and needs out of the discourse. However, a belief in objectivity and detachment does not keep an individual from being highly sensitive and affected over time by the surrounding institutional climate, but it may prevent the individual from acknowledging and compensating for these influences.

Attitudes Toward National Security, Defense Work, Citizenship, and Social Responsibility

Engineers' attitudes toward conversion have not been specifically examined on a significant scale. But some predictions can be made on the basis of a number of interrelated factors that have been studied, albeit incompletely: their attitudes toward national security and defense work and the ways these may change in our highly fluid international scene; the role of the military in national security; their judgment of the technology needed for effective defense; and their capacity to be mobilized in the workplace, in professional and civic organizations, and at the ballot box.

One of the most complex, significant, and underresearched questions is the concept of professional responsibility that is promoted by engineering education and professional organizations. In general, engineers are seen and taught to see themselves as servants of a social consensus whose contribution requires a detached objectivity. As Florman writes, "It is not the engineer's job, *in his or her daily work*, to second-guess prevailing standards of safety or pollution control, nor to challenge democratically established public policy. As a citizen, to be sure, he may work for modification of such standards and policies."[21]

As noted earlier, sociopolitical values are not the primary factor behind most engineers' choice and enjoyment of their jobs. This is in keeping with the value engineering education places on keeping personal beliefs and feelings out of workplace decision making. The modest membership of engineers in organizations formed to address their social contribution—from High Frontier to American Engineers for Social Responsibility—suggest that this education succeeds.

Several authors have observed that engineering is dominated by generally conservative professional organizations.[22] But the notion of conservatism is imprecise and fluid, and so is the degree to which ideology actually intrudes into these organizations' activities. Moreover, since only one-third of engineers belong to those organizations, their ideological influence on the profession as a whole is unclear.[23]

The engineering population is generally seen as politically quiescent and shunning extreme positions. Zussman observes that two-thirds of a sample of engineers said they had engaged in political or civic activity beyond voting, but a majority of these instances were signing a petition, attending a town meeting, and similar activities that did not involve high commitment or risk. The engineers most likely to participate in political or civic activity are those whose lives are relatively stable and who work close to home.[24]

In spite of this general pattern, engineers, like scientists, have been dramatically mobilized at times in history. Layton's *The Revolt of the Engineers* describes a rather sedate revolution motivated by enlightened self-interest: a desire for recognition and more favorable working conditions.[25] Peter Kuznick's *Beyond the Laboratory: Scientists as Political Activists in the 1930's* reveals a more profound shift in the science and engineering cultures in an earlier era of swelling grass roots activism: "A remarkable transformation occurred in the inner world of scientists [defined to include engineers]—a change in the beliefs, attitudes, and assumptions that defined their social identity and sense of purpose—a change more profound than that experienced by any other sector of the American population."[26]

No major engineering society has surveyed its members in any detail on their attitudes about national security, military contracting, or priorities for the technical community since the beginnings of Soviet glasnost and perestroika. But the available empirical data, drawn from the mid-1980s and earlier, show a mix of sentiments. For example, a 1984 survey by the Institute of Electrical and Electronics Engineers found that 39% of responding members would prefer not to engage in military work and 48% would prefer not to work on nuclear weapons. On the other hand, 56% of respondents said they would be proud to work on weaponry, and when the subject was nuclear weaponry that percentage only fell to 42%.[27]

This suggests that the profession is unlikely to reach a consensus for conversion initiatives. If conversion is undertaken, driven by a strong enough national or corporate mandate, a willing labor pool of engineers will be available. But without strong leadership, this very mix of sentiments, coupled with many technical people's decision-making styles (as discussed earlier), may lead engineers to support the status quo in the interest of harmony or predictability.

The numeric balance of "pro" and "anti" does not address how ambivalent engineers feel about military work. Several observations, covering more than engineers but applicable to them, suggest that attachment to weapons work (or any technology) can be powerful.

The literature of technical forecasting, for example, provides evidence for this view. In *Megamistakes*, a study of three decades' worth of technological forecasts in the popular press, Steven Schnaars evaluates engineers' and scientists' predictions of technical innovations.[28] He cites Avison and Nettler's 1976 finding of a "strong unwarranted optimism in forecasts" and observes that "most forecasts fail because the forecasters fall in love with the technology they are based on and ignore the market the technology is intended to serve."[29] These findings fit the common experience of emotional investment in projects under way in any field.

Military technologies—especially nuclear weapons and "smart" missile systems—fascinate many technical people. Thomas Grissom, a physicist at Sandia National Laboratories who supervised a group responsible for making triggering devices for nuclear weapons, observes that "when most people first see a nuclear explosion, their instinctive response isn't terror or revulsion. It's fascination with the big boom and the light."[30]

One theory to explain the predominance of enthusiasm over skepticism is advanced by Carol Cohn, who has studied the language and thought of nuclear defense intellectuals—some of whom are engineers, all of whom shape the cultures in which defense engineers work. In "Sex and Death in the Rational World of Defense Intellectuals," she notes in their use of language a highly charged emotional imagery coupled with an adamant assertion of detachment and rationality.[31] This combination is maintained by a number of linguistic feats. One in particular stands out. In the discourse of professionals who create—and advocate—new military technologies, the reference point is rarely the human beings potentially affected by these weapons. More often the weapons themselves are anthropomorphized. Thus a "survivable" nuclear weapon is one that is likely to be preserved in its silo during an attack, that is, a weapon that threatens the survival of thousands or even millions of people.

Anthropomorphizing weapons technology has continued unabated, as does the reverse process, dehumanizing humans.   On the current infatuation with "smart" weapons, Cohn writes,

> The language of warfare has always been abstract and euphemistic, but the Gulf war has apparently inspired briefers to new heights of obfuscation.   Listening to the news, you would hardly know that U.S. forces are bombing Iraq and Kuwait.   It appears that the only things that get "killed" on these sorties are Iraqi missiles and other military targets.   Iraqi people do not get killed; civilian casualties are referred to as "collateral damage"—a stunningly abstract and sanitized way to refer to mangled human bodies.   The word "collateral" serves to remove moral responsibility from the attacker; the deaths are "collateral," secondary, not at all what we intended.   The word "damage" serves to turn human beings into objects: things are damaged; living beings are hurt.
>
> Instead of humans, weapons become the living things and the "vulnerable" actors in warfare.   Weapons take on human attributes.   Bombs are either "dumb" or "smart."   "Smart" weapons have "eyes" and computer "brains."   The computer "makes a decision" when to drop seven and a half tons of bombs. The moral responsibility of the combatants disappears.[32]

Cohn found that when she tried to raise these issues with defense intellectuals—even those who had worked closely with her and initially welcomed her—she was immediately seen as naive and marginalized from the group.   The same experience has befallen many who have tried to interest engineers and scientists in economic conversion on ethical grounds.

We do not know how widespread or consistent these patterns are or how significant they are in the motivational systems of engineers who work in weapons industries.   But one additional observation makes them especially worth exploring.   Cohn, whose general perspective is critical of warfare and whose writing is based on a year of intensive field research, reports that, with exposure to her subjects, her own patterns of thought began to change.   "I started this research because I couldn't understand how they could think like that.   But eventually, my question became, *How could I think like that?*"

Cohn gives us striking evidence that group psychology is as critical as individual psychology in understanding engineers' relationship to military work and, therefore, to conversion.   As noted earlier, individually engineers differ widely in their social values and in the relationship of social values to professional ethos.   But groups of engineers tend to be remarkably homogeneous in these respects.   William Broad's *Star Warriors*, about the Lawrence Livermore Laboratory, and M. Grace Mojtubai's *Blessed Assurance*, about the Pantex nuclear weapons assembly plant in Texas, portray two radically different sets of sensibilities.   The first community shares a zeal for work and technical achievement, irreverence, and even jadedness about the ways of the world, and a taste for "black humor."   The second community is characterized by respect for authority, a highly polarized worldview, and a version of fundamentalist Christianity, which for many includes the view that nuclear weapons will be part of the apocalyptic battle between good and evil.   These cultures are dramatic in their differences, but within each culture highly consistent.

Internal consistency suggests that, beyond engineers' private beliefs about national security and work ethics, responses to economic conversion or any other radical change are more likely to be shaped by the dynamics of the groups in which they function: departments, organizations, professions, or industries.   The importance of group dynamics and organizational cultures is noted elsewhere.   In a study of crisis-prone organizations, Ian Mitroff notes a common feature of "bounded emotionality" in decision makers and corporate cultures—a tendency toward simplified, depersonalized discourse and poor intrapersonal skills, i.e., the ability to recognize and communicate one's own emotion and bias.[33]

A growing body of qualitative research on the cultures, group dynamics, and patterns of discourse of military-industrial cultures suggests engineers' viewpoint is an important area for exploration.   For example, in *Minds at War,* Steven Kull lists a number of ways in which nuclear policymakers and opinion leaders selectively apply their knowledge about the consequences of the weaponry.[34]   These include denial, suppression, and rationalization.   Coupled with the finding that engineers tend to internalize the value systems of their organizational cultures, this observation takes on greater significance.   In "Resistances to Knowing in the Nuclear Age," John E. Mack notes that denial, suppression, and rationalization take place on two levels: individual resistances, "which can be overcome, at least to a degree that permits involvement, by psychological work that reduces the margin between intellectual and emotional knowing" and collective resistances "at the level of the society or nation as a whole and its organizations and institutions."

In psychoanalytic terms, Mack continues,

> To know in the fullest sense the nature of the destructive monster we have created delivers a kind of double blow to our experience of reality and ourselves.   On the one hand, it confronts us with the failure of our political and military system to order its affairs in a way that provides security for its people in relation to other nations.   At the same time,

such knowledge brings us face to face with the fact that our policy of a "credible" nuclear deterrent requires the willingness to annihilate most, if not all, of human life in the service of national goals—a piece of societal self-knowledge that is potentially devastating to our collective self-regard.[35]

This is obviously a value-laden statement; the impact of the arms race on individuals' self-regard depends to some extent on their beliefs about international security. However, questions of the internalization and rationalizing of what our military does are part of the "baggage" of the conversion controversy, and we are unlikely to resolve such questions until we confront these issues both as engineers and civilians.

Toward a Psychology of Empowerment

Given engineers' tendency to defer to management on questions of policy, motivating engineers in support of economic conversion requires some combination of (1) motivating recognized leaders and role models and (2) helping to bolster engineers' sense of interpersonal efficacy and social responsibility. Without at least one of these changes, proposals for conversion might meet the greatest resistance, ironically though it seems, from those engineers who are the most ambivalent about military work, but who have internalized the value systems of their organizations as a coping mechanism.

More specifically, what kinds of participation and leadership will be required of engineers? They will need to:

(1) cope with institutional uncertainties and maintain morale around them;
(2) develop alternate products and processes for military-based industries, teach and motivate their peers to acquire necessary new skills;
(3) use their power in the labor market and raise their voice in professional societies to support converted and converting industries and encourage openness to the idea in defense-dependent companies;
(4) vote for candidates who support conversion legislation, help to get the issue onto the political agenda, and legitimate it with their seal of approval as socially responsible professionals;
(5) accept changes in some cases, in career path and loss of "perks" as ties with government contracting loosen.

Several bodies of psychosocial literature provide insight into enacting these contributions:

(1) determinants of voting and political behavior;
(2) leadership studies;
(3) determinants of antinuclear and/or antiweapons activism;
(4) determinants of social/environmental activism.

This body of literature is especially important since the environmental crisis is one of the major competitors for federal funds and national attention presently devoted to military defense. These three areas are converging into a new and incomplete field, sometimes known as the psychology of empowerment. A brief review of research, which is mostly qualitative and based on observations and testimonies of activists, suggests the field is just beginning.

On Political Participation

Of the rapidly evolving field of political psychology, three areas are of special relevance: values and attitudes; political participation in local and national elections, in professional organizations, and in workplace "politics"; and the relationship of cognitions, emotions, and unconscious dynamics in shaping political choices.

The far-reaching interview study, *Habits of the Heart,* offers insight into the individualist political culture of which engineers are a part.[36] In it, Robert Bellah and his coauthors note that the term "politics" carries three fundamentally different meanings and that participation differs with these operational definitions. Politics can mean "making operative the consensus of the community, reached in free, face-to-face discussion; or the pursuit of differing interests according to agreed-upon, neutral rules; or 'the politics of the nation,' which exalts politics into the realm of statesmanship in which the high affairs of national life transcend particular interests." The first and third definitions—the politics of community and nation—are favorably regarded by most Americans, while the politics of interest groups are disdained. Certainly the economic conversion debate can be framed at any of these levels: as a process by which established groups in the workplace engage in communal soul-searching about their activities and purpose; as a highly lopsided tug-of-war between the Pentagon and civilian markets for contractors' loyalties; or as a transcendent debate about what really constitutes national security. If engineers' hearts follow the same habits as the general population, then we should expect that conversion would win over the most hearts and minds if framed as a unifying realignment of purpose of company or the nation rather than as bashing the Pentagon.

While engineers have the capacity for political activism and display the full range of political viewpoints, it is safe to say that most engineers are not activists themselves. The literature on political behavior of marginal voters therefore has some relevance to mobilizing engineers' votes. Cedric Herring offers useful findings on the motivational patterns of politically alienated adults:

(1) those with organizational affiliations and trust in others are less likely to be "political dropouts."
(2) those who feel worse off or underemployed are less likely to be "political ritualists."
(3) those with organizational affiliations are more likely to engage simultaneously in conventional and unconventional modes of participation.
(4) those who are untrusting and those who feel underemployed are significantly more likely to be involved as protesters.[37]

Again, here is evidence for the need to mobilize not only individuals, but coherent, sustainable group efforts. Not only a rational plan, but strong interpersonal and intrapersonal skills are essential for success.

On Political and Workplace Activism

In most conversion scenarios, a majority of engineers will be called upon to exercise not leadership but "followership"—the willingness to work with an evolving group consensus, cope with change, offer ideas, and learn new skills. This involves an important psychological orientation, sometimes called prosocial behavior. Prosocial behavior is about cooperation, conscientiousness, and responsibility in a rather conventional sense. An engineer who receives a memo from the boss inviting suggestions for diversifying into civilian business, and who devotes some evenings and weekends into developing proposals, is exercising prosocial behavior. Alfie Kohn's *The Brighter Side of Human Nature* is a recent and comprehensive review of the roots of prosocial behavior.[38] In another relevant study, Ervin Staub focuses on three simple elements of prosocial behavior: a positive view of people in general; concern about the welfare of others; and an attitude of personal responsibility.[39]

In addition to these prosocial qualities, however, some engineers will have to be leaders for conversion to succeed. Reasons for this are, first, initiatives for change will be accepted most easily by rank-and-file engineers when they come from respected peers rather than exclusively from management or politicians. Secondly, the input of engineers regarding questions of technical feasibility and of policy decisions is essential in cost-effective conversion strategies. Policymakers on their own will not always know the questions that must be asked. Finally, the political quiescence of engineers and their communities' role as voting blocs in support of defense spending has been a factor of unknown significance in impeding conversion to date. Visible, trustworthy grass roots leadership will be required to break this cycle. Perhaps the best summary of the vast literature on leadership is Warren Bennis' observations that "the best leaders are self-evolvers" and "the process of becoming a leader is much the same as the process of becoming an integrated human being."[40]

An essential element in leadership is breaking away from the influence of the group and learning to dissent. Much of the literature on dissidence focuses on human behavior in extreme situations, but it is applicable wherever human beings face stress and change. In *Crimes of Obedience*, Herbert Kelman and Lee Hamilton compare the inner worlds of people in two diametrically opposed cultures.[41] One is the French village of Le Chambon, where in World War II 4000 villagers hid 4000 Jews from the Nazis and suffered only a handful of casualties. The other is the U.S. army company that performed the My Lai massacre in Vietnam. The first of these cultures was one of risk-taking for the greater good, discipline, creativity, and long-range vision. The second was the quintessential "me-first" culture: violent, reactive, unable to see past the moment. Both cultures were forms of "groupthink," in which individuals were immersed in the momentum of group action. But in one, the group dynamic served to enhance individual responsibility and creativity, whereas in the other, these qualities were destroyed.

The differences between the moral discourses of the two villages were simple and profound. The French villagers showed three qualities: a sense of independent moral agency or the perception that one is capable of making significant choices; an attitude of caring for others; and an appreciation of causal relationships between actions and consequences. The American troops lacked these qualities; interestingly, these same qualities are ones that engineers are encouraged to divorce, to a great extent, from their "professional" thinking.

Two overarching psychological frameworks are worth noting. One is Abraham Maslow's notion of the hierarchy of needs. By this theory, people are able to look at long-term issues when their immediate needs are met and able to address needs perceived as secondary, such as job satisfaction or health of the economy, only when primary needs or survival, whatever that might mean to a particular individual, are met.

Another important factor is Erik Erikson's research on the human life cycle, which suggests that different motives dominate individuals at different points in their lives. For example, issues of identity and security are most important in the early years of establishing oneself in a career. Seen in this light, the fact that entry-level engineers overwhelmingly affirm their desire to earn high salaries does not necessarily mean this is a lifelong incentive. By midlife, many engineers may well show the common human tendency to be interested in the well-being of others and seek a way to help future generations. Florman observes, "When older engineers get together they invariably agree that immediately after graduating from college they wished they had taken more technical courses. Ten years later, advancing along career paths, they wished they had learned more about business and economics. Ten years again, in their forties, thinking about the nature of leadership and musing about the meaning of life, they regretted not having studied literature, history, and philosophy. This pattern has become something of a cliché, confirmed by studies and polls."[42] Might it be, then, that younger engineers would be most likely to be enthusiastic about conversion if it were packaged as a source of innovative startup opportunities with high potential for recognition and reward, while midcareer engineers would be more likely to be motivated by the hope of leaving a healthy economy or a more peaceful and stable world for their children?

Based on literature review and observation of several groups of students in a new course, "The Psychology of Global Awareness and Social Responsibility," clinical psychologist Sarah Conn proposes a four-level model of the development of "global awareness and social responsibility":

(1) cognitive (knowledge of the issues);
(2) affective (emotional response to the situation);
(3) connection to the issues in a personal manner; and
(4) action (which reinforces one's sense of efficacy and identification with the issue).[43]

Many writers have expanded on these areas in recent years. For example, Chellis Glendinning points to the importance of such elements as tolerance of ambiguity, the acknowledgment of negative visions to allow the creation of positive visions; committing consciously; and need expressing psychological truths.[44] Mary Watkins adds the importance of dreams and the subconscious in weaving together dissonant elements of the self in the interest of overcoming paralyzing conflicts.[45] Finally, Doug MacKenzie-Mohr has found that the major difference between two groups of students concerned about nuclear issues in which one was activist and one not was in their perceived tactical efficacy of possible actions.[46] That is, those who acted were those who could think of (or were shown) something useful to do.

At any given time, it seems, a minority of people in any field have what it takes to be activists: the luxury of introspection, the access to information, the peer supports, and the awareness of effective paths of action. Based on what we know about engineers, where might we find those activists in the profession and how might we create a few more?

In light of engineers' strong need to see themselves as social contributors and their weak attraction to rebellion, conversion is probably best portrayed as a move toward emerging fields of research, development, and production, such as environmental cleanup, communication, transportation, and biomedical applications issues integral to security in an interdependent world.

Engineers will be most receptive to conversion initiatives that take their personalities and information-processing styles into account. For example, the tendency to believe in absolute truth may make engineers hesitant to enter complex, value-laden debates until some personal experience or compelling argument is sufficiently galvanizing. Then they may be zealous advocates, but may also be disaffected as suddenly and absolutely as they were mobilized.

The characteristic utilitarian worldview of engineers may be both a blessing and a barrier. Conversion advocates must be able to justify the disruption of routines that will accompany major institutional change. But if a case for the alternative products can be made compelling enough, and if the institutional mandate is expressed strongly enough so that adaptation is seen as responsible professional behavior, then the professionalism and task-orientation of engineers may be turned to an advantage.

Using Conn's model of the four dimensions, motivating engineers to act is likely to be the easiest goal to achieve. Once engineers are convinced that a path is useful, they will not be handwringers, but will want to get moving. The other dimensions, which involve engaging engineers cognitively, will be demanding but possible—*if* unacknowledged emotional and subconscious blocks such as externalization of responsibility and fascination with weapons technology can be addressed. The most difficult but necessary challenge will be the emotional dimension. For all four dimensions, the stereotypes of engineers and the polarizing style of many activist campaigns may make the problem look more insurmountable than it need be.

To encourage the creative involvement of as many engineers as possible in economic conversion efforts and to draw on all possible dynamic leadership in the engineering ranks, the conversion process should be designed to be sensitive to their needs, grounded in an understanding of their motives, and planned to make good use of their strengths. More specifically:

(1) where possible, motivate and involve small groups rather than individuals so that peer support is available and no individual is forced to play the role of the isolated dissident;
(2) use models of success that are as similar as possible to a given group of engineers' own environment;
(3) identify and publicize cutting-edge technological challenges in the civilian sector and create incentives for applying engineering talent to these;
(4) model the process on other scientific mobilizations in the public interest, such as environmental technologies, the space program, and AIDS research;
(5) respect (and perhaps use) engineers' need for order and a sense of control;

(6) find arguments that are convincing to engineers about the need for conversion and its likely rewards;

(7) understand that many engineers already feel challenged, accused, and besieged and therefore concentrate on the benefits of conversion for the future, rather than implicitly repudiating the work engineers have done in the recent past.

Perhaps the most useful model of the engineer-activist for economic conversion is not the militant figure refusing to work on weapons, nor the lone whistleblower facing the media, but the inventor in the basement who dedicates years of quiet creativity to making a new idea work. Engineers are most likely to get involved in conversion efforts if they can do so comfortably, work in harmony with their peers, and see practical results from their efforts. The enormous attractiveness of high-technology startups in the 1980s shows that engineers are willing to take risks and can thrive on chaos, given the right systems of incentives and supports.

In examining the potential for change in public controversies involving science and technology, a major source of creativity is in the young field of risk communication. This is the study of factors that make a difference in how a risk is perceived and thus in whether people fight or accept it.[47] These include whether the risk is voluntary or imposed, whether it is natural or created by humans, and whether it is dreaded or not. This body of knowledge is really a broad discourse about human beings' responses to threat, change, and each other. The usefulness of this discourse is illustrated by the work of Peter Sandman, head of the Environmental Communication Institute at Rutgers, who entertains both corporate and activist audiences with the parable of "The Technocrat and the Housewife."

> She comes to express outrage because an unknown hazard has been imposed on her. He answers with patient, rational explanation that the hazard is minimal. Because he hasn't heard what she came to communicate—her outrage—she repeats it more emphatically. Because his reassurance hasn't worked, he retreats into his charts and graphs. Finally, she is waving her leukemic child in his face and he is staring cataleptically at his printout.

No one has yet applied this difference in perception to the economic conversion debate. Framed this way, the focal question becomes: why do so many engineers and others in the technical community view the risks of defense work—from macroeconomic uncertainty to workplace safety—as acceptable while appearing to view the risks of conversion as unacceptable?

Sandman, who has been extensively involved in educating both sides in corporate-community debates, believes technical professionals possess a great deal of untapped potential for dealing with the human dimensions of controversies—if only an authority they respect gives them permission to do so. He says, "I tell engineers they have been making a category error in the way they deal with the issues—public sentiment is not a technical system to be managed; it's a human system. I ask them to imagine how they would respond if their daughter came home from school in tears because she had been attacked. For the most part, given permission, engineers can be as empathic as anyone else."

This vignette—and especially the use of the term "category error"—illustrates the monumental importance of the final element in Conn's model, the dimension of connection. Motivating engineers to constructive involvement in conversion efforts—either as leaders or as followers—requires that they build bridges of values and meaning to their work. Anthropologist Lisa Peattie considers economic conversion to be "a set of organizing ideas" involving not only economic and technical issues but people's fundamental motivations and systems of meaning.[48] What will it take to formulate the conversion agenda that brings alive the alternatives and reasons for change yet reaches engineers in their own language?"

Here is a speculative list of social and economic trends that could be capitalized upon to strengthen the case for conversion:

1. Civilian challenges in science and technology. Ted Taylor, the nuclear physicist who abandoned work on weapons and eventually on power plants, now works on problems like hydrogen fuels and ocean water desalination and says, "I have all the excitement and challenge I could want. It's a myth that you have to work on weapons to do interesting work."[49] Environmental cleanup and "clean technologies" also offer enormous technical challenge and economic opportunity, and this has become a focus for some state-level economic development programs, for example, Massachusetts.
2. Sustainable economic development—not only revitalizing our own economy but working in the eastern bloc and the Third World. An extremely useful model for "selling" conversion is the proceedings of a conference sponsored by the Institute for Defense and Disarmament Studies in 1990, "Swords into Ploughshares," which brought together spokespeople from the U.S. and Soviet Union's high-technology industry for three days of enthusiastic discussion about civilian joint ventures.[50] In light of the spectacular changes in that region since then, there is likely to be a ground swell of interest in "making deals, not war."
3. Civilian ventures in outer space, including joint ventures with other nations. The militarization of space and of the space program makes this a charged issue, but its possible value lies in the shifts in consciousness. Even non-space-travelers are led to imagine the earth as a whole by space travel. As astronaut Rusty Schweikart says,

When you go around the earth in an hour and a half, you begin to recognize that your identity is with that whole thing. That makes a change. You look down and you can't imagine how many borders and boundaries you cross. Below you, hundreds of thousands of people are killing each other over some imaginary lines you aren't even aware of. From where you are, the planet is whole, and it's so beautiful that you wish you could take each individual by the hand and say, "Look at it from this perspective. Look at what's important."[51]

4. Changes under way in corporate structures and cultures based on new management principles, many of which can empower working people and make holistic thinking possible. Two examples:
    (a) At Stanford Graduate School of Management, Michael Ray teaches a popular course called "Creativity in Business." Students cultivate their creativity (a permissible value by conventional standards) through unconventional methods: meditation, journal writing, art therapy, wilderness vision quests.
    (b) Peter Senge's *The Fifth Discipline*, a popular new book uses principles of cybernetics to call for new modes of thinking in management and industry.
5. The interaction of U.S. industry with its Asian counterparts is producing a "Pacific shift" that is not only economic and geostrategic but cultural as well.[52] As western companies develop working relationships with people for whom Japanese management and eastern religions are not exotic concepts, cross-cultural understanding will become more and more essential to doing business productively. The pace of technological change and social breakdown may make us ripe for learning the wisdom these cultures have to offer.

Conclusions

Roles and Responses

The impact of unplanned economic conversion on engineers—a process involving job loss and large-scale uncertainty—would be moderately to very traumatic. Planned economic conversion would still involve adjustments, but could quickly show benefits, depending on how the process is managed and motivated.

Engineers' impact on a process of planned conversion could range from cooperation and enthusiasm to disinterest or even resistance. Again, the enormous variation depends on the scenario. In any plausible scenario, the extremes of large-scale, dynamic advocacy and open rebellion are unlikely.

Engineers can be thought of as a "swing vote" in deciding whether, and how coherently, economic conversion will take place. Engineers, like other people, oscillate within a range of attitude and behavioral possibilities. They can be renaissance people—or drones. They can be rigorously principled—or corrupt. They can be a force for social change or for the status quo.

To prevent engineers from experiencing the negative consequences of resistance and discomfort, economic conversion should account for their experiences and interpersonal styles.

Implications for Policy

The personality tendencies, decision-making styles, and work experience of engineers have a number of implications for economic conversion policy. First, consider the leadership and the management of change. In his pivotal essay, "Pacific Shift," William Irwin Thompson points out the paradox that "the cumulative effect of electronics and aerospace technologies is a dislocation, a deconstruction of civilization," but "the kids and the philosophers understand this better than the technicians and the managers, for those in the middle do not see the edges." Common experience of workplace frustration, messages of inadequacy, and diminished expectations over time suggest that simply promoting dissatisfaction with things as they are will not inspire initiatives toward change but is more likely to lead to resignation. The "carrots" of positive alternatives, defined as things attainable, are more useful as incentives than an analysis of the weaknesses of a militarized economy.

Secondly, the group cohesion of technical people, coupled with the reluctance or inability to take authority on policy issues, suggests that incentives and plans for change are best focused on such groups as, for example, department, professional specialty, and sometimes company, rather than on isolated individuals or monolithic large groups. These group dynamics also suggest that peers are the best advocates for changes in direction, for instance, executives of successful converted businesses will be the best sales force for the idea of conversion to other executives; engineers who have retrained and hold satisfying civilian jobs will be the best role models for other engineers.

Most engineers can hold the conversion debate with its moral, psychological, social, and cultural dimensions at arm's length. By emphasizing technical and economic issues, the necessary changes appear safe for the technical community. But the process, if carried out by ignoring key elements, may fail to achieve much of its potential for economic and institutional revitalization. Economic choices reflect beliefs about self and world, whether they are articulated or even acknowledged. For a generation, we have defined security in terms of national inviolability and sought that security by means of weapons of mass destruction. A redefinition is under way, toward a view of security based on international relationships coupled with social and environmental well-being. That transformation can only proceed on the macro level if accompanied by individuals' self-conscious, purposeful exploration of their sense of self, motives for working, and the social impact of that work.

Implications for Engineering Education

In these times of change, as we all begin to understand the interconnectedness between technical and policy questions, engineers must be taught a radically new concept of professional responsibility. The ideas of detachment and objectivity remain important when applied to physical systems. But, in their inevitable dealings with social systems, engineers must learn to think systematically on a new level, appreciating the human and environmental impact of their technical decisions and applying their full human wisdom to the value questions that are inseparable from their work.

Efforts to add humanities and arts to engineering curricula may help marginally in cultivating these abilities. But the greater, unrecognized need is for experiential education that will build interpersonal and intrapersonal skills. These include psychological literacy and self-knowledge. They also include empathy and a capacity for communication that is not only clear but as complex as the world it represents. In short, education of engineers must change radically to make room for not only awareness of the world but also of the profound human factors that are intrinsic to decision making. This will probably happen most effectively in climates that are as free as possible from military research and development funding. One of the most powerful catalysts will be positive role models and successful cases.

Finally, engineering education must place a priority on equipping students with a knowledge of the choices before them. Even in universities where military funding shapes the research agenda, students must learn about the civilian areas in which their expertise may be used. Career services offices could contribute enormously by teaching students how to identify and locate promising civilian job opportunities.

Implications for Work with Professional Societies

The crisis of the military sector may offer opportunities to engineering professional societies. It provides an enormous incentive for self-knowledge and for service, that is, for surveying the attitudes and needs of members and thinking creatively about ways to address those needs. Engineering societies, like engineers, are likely to be better motivated by carrots than by sticks. Helping their members to meet the challenge of conversion means making a quantum leap in social relevance and, potentially, growing in membership as a result.

Alliance with mainstream groups is likely to be most effective when it:

1. builds on existing trends and addresses issues perceived as important, such as:
   (a) Why are ethics and human values such perennial concerns in engineering and engineering education and yet so hard to teach?
   (b) Why is there so much tension between technical communities and the public on critical policy questions including, but not limited to, defense policy?
   (c) How can the profession take care of its own in times of economic chaos and dislocation?
   (d) How can we learn more about the hearts and minds of our members, in order to serve them and therefore keep them?
2. emphasizes positive initiatives rather than traumatic adjustment;
3. provides concrete resources of use to professional societies, such as speakers, programs, publicity and feedback.

These aims might be achieved by launching a joint research initiative, perhaps based in a university and cosponsored by the major engineering societies, to assess engineers' attitudes about defense work and conversion and to develop concrete services based on the findings. These services might include workshops to help members update their skills, clarify their values and interests, and research the job market.

Research Agenda

The gaps in our knowledge based on these themes are big enough to drive a B2 bomber through. The existing research on the psychology of engineers is limited by the following characteristics:

1. small samples;
2. outdatedness in light of changes in the engineering population and public perceptions of national security issues;
3. lack of accounting for differences based on professional specialty, education, corporate culture, age, gender, and other personal variables;
4. inconsistent theoretical frameworks.

Further research is critically important and time-dependent. The best strategy for meeting this need would be a multidisciplinary research initiative supported by professional societies and/or major defense employers. It would include broad and detailed survey research on attitudes about national security, economic conversion, and sources of job satisfaction. It would explore the variations in engineers' responses to different conversion scenarios. Such a study would be complemented by more sophisticated exploration of constructs of social responsibility, their political behavior and its determinants, and the factors underlying their choices regarding defense employment and its alternatives.

A laundry list of focusing questions includes these:

1. How do engineers' attitudes about military work and personal responsibility vary with job experience, industry, point in life cycle, political climate, institutional climate, and (many) other variables?
2. To what extent do engineers' attitudes about national security and professional ethics stem from innate personality characteristics, and to what extent are they due to work experience?
3. To what extent do the psychological phenomena found in intellectuals studying defense and generally in military-industrial cultures, such as fascination with weapons technology and denial or minimization of its human impact, apply to engineers?
4. What do most engineers believe about economic conversion, and how have their beliefs been defined?
5. What are commercially viable civilian technologies with the potential to fascinate engineers and be useful to society?

There are two approaches to answering these questions. The first is the survey or interview study. The second is to examine actual cases of conversion that are underway. The first approach offers more generalizable data, but the second offers greater insight into the applicability of any conclusions to particular conditions. An added benefit to monitoring conversion efforts and offering feedback to those involved in making them work is that the feedback they receive can be a powerful tool for adjustment.

This has been an eclectic survey of the known universe of information about engineers and about determinants of the behaviors required for their constructive participation in economic conversion. It reflects the incompleteness of our theoretical and empirical knowledge and affirms a need for significant research. This research is most likely to succeed in the context of an action agenda—one designed to engage engineers with the personal involvement and creativity that the profession sees as its hallmark.

Notes

[1]Samuel Florman, *The Civilized Engineer* (New York: St. Martin's, 1987), pp. 76-77.

[2] G. Haworth, R. Povey, and S. Clift, "The Attitudes Toward Women Scale (AWS B): A Comparison of Women in Engineering and Traditional Occupations with Male Engineers," *British Journal of Social Psychology* 25 (1986): 329-334.

[3]Henry H. Bauer, "Against Interdisciplinarity: Implications for Studies of Science, Technology, and Society," *Science, Technology, and Human Values* 15, No. 1 (1990): 105-119.

[4] W. Grogan, "Engineering's Silent Crisis," *Science* 247 (26 January 1990): 4941.; E. Layton, *The Revolt of the Engineers: Social Responsibility and the American Engineering Profession* (Cleveland: Case Western Reserve University, 1971).

[5] M. Larson, *The Rise of Professionalism: A Sociological Analysis* (Berkeley, CA: University of California Press, 1977), p. 30; P. Meiksins, "Professionalism and Conflict: The Case of the American Association of Engineers," *Journal of Social History* 19, No.3 (1986): 403-421.

[6] R. Zussman, *Mechanics of the Middle Class: Work and Politics Among American Engineers* (Berkeley, CA: University of California Press, 1977): pp. 102-123.

[7] Linda M. Donald, "Behavioral Styles of Engineers in a Research and Development Organization," M.S. thesis from National Technical Information Service, U.S. Department of Commerce, (UCRL-53898, Distribution Category UC-700).

[8]The Myers-Briggs test, based on Jungian principles is described in depth in Isabella Briggs and Peter B. Myers, *Gifts Differing* (Palo Alto, CA: Consulting Psychologists Press, 1980). The test is based on the idea that knowing one's style allows one to accentuate strengths and compensate for differences with others.

[9] Jerald M. Henderson, Laura E. Bellman, and Burford J. Furman, "A Case for Teaching Engineers with Cases," *Engineering Education*, (January 1983): 288.

[10] *New York Times*, "Survey Finds Materialism Rising Among College Freshmen" (14 January 1985). The extent of this mood swing is evident when one opens Perucci and Gerstl's (1969) The Engineers and the Social System, optimistically dedicated "to the new technocrat, whose humanity will guide the use of his talents."

[11] Walter H. Griggs and Susan L. Manning, "Money Isn't the Best Tool for Motivating Professionals," *Personnel Administrator* (June 1985): 63; Additional research now underway by Bailyn et al at the Sloan School of Management applies self-efficacy theory to the question of engineers' career anchors.

[12] James Watson and Peter Meiksins, "What Do Engineers Want: Work Values, Job Rewards, and Job Satisfaction," *Science, Technology, and Human Values* 16, No. 2 (1991): 140-172.

[13] Susanne D. Ellis, *Physics Today* (1986): 84.

[14] Aaron Bernick and LaBier are quoted by Ramon G. McLeod and Allyn Stone, "Baby Boomers Want Meaning," *San Francisco Chronicle* (13 February 1989): A4.

[15] J. Pfeffer, "A Partial Test of the Social Information Processing Model of Job Attitudes," *Human Relations* 33, No. 7 (1980): 457-476.

[16] L. Bailyn and J. Lynch, "Engineering as a Life-Long Career: Its Meaning, Its Satisfactions, Its Difficulties," *Journal of Occupational Behavior* 4 (1983): 263-283; F. Guterl, "Spectrum-Harris Poll: The Career," *IEEE Spectrum* 21 (1984): 59-63.

[17] K. Alexander, "Scientists, Engineers, and the Organization of Work," *American Journal of Economics and Sociology* 40, No. 1 (1981): 51-66.

[18] R. Kopelman, "Psychological Stages of Careers in Engineering: An Expectancy Theory," *Journal of Vocational Behavior* 10, No. 3 (1977): 270-286.

[19] Kandebo Grogan et al. "U.S. Faces Potential Shortage of Engineers," *Aviation Week and Space Technology* (5 December 1988): 36-38; William Booth, "Engineers Hear a Competitive Parable," *Science* 238 (1987): 474.

[20] Kandebo, 36-38; Florman, 76-77.

[21] Florman, 99.

[22] Layton; Meiksins, "Professionalism and Conflict," 403-421.

[23] Florman.

[24] Zussman.

[25] Layton.

[26] Peter Kuznik, *Beyond the Laboratory: Scientists as Political Activists in 1930s America* (Chicago: University of Chicago Press, 1990), p. 253.

[27] IEEE, "Readers Express Mixed Attitudes on Personal Involvement in Defense Work," *IEEE The Institute* (February 1984): 3.

[28] Steven Schnaars, *Megamistakes* (New York: Free Press, 1989), pp. 9-10.

[29] William Avison and Gwynn Nettler, "World Views and Crystal Balls," *Futures* 8, No. 1 (February 1976).

[30] Melissa Everett, *Breaking Ranks* (Philadelphia: Society Publishers, 1989).

[31] Carol E. Cohn, "Sex and Death in the Rational World of Defense Intellectuals" 12, No. 4 (Fall 1987): 687-718.

[32] Carol E. Cohn, "Decoding Military Newspeak," *Ms.* (March/April 1991): 88.

[33] Ian I. Mitroff and Thierry C. Pauchant, *We're So Big and Powerful Nothing Bad Can Happen to Us* (New York: Carol Publishing Group, 1990), p. XIII.

[34]Steven Kull, *Minds of War* (New York: Basic Books, 1988), pp. 296-318.

[35]John E. Mack, "Resistances to Knowing in the Nuclear Age," *Harvard Education Review* 54, No. 3 (August 1984): 260-270.

[36]Robert N. Bellah, Richard Madsen, William M. Sullivan, Ann Swindler, and Steven M. Tipton, *Habits of the Heart: Individualism and Commitment in American Life* (Berkeley, CA: University of California Press, 1985), pp. 200-203.

[37]Cedric Herring, "Acquiescence or Activism? Political Behavior Among the Politically Alienated," *Political Psychology* 10, No. 1 (1989): 135-151.

[38]Alfie Kohn, *The Brighter Side of Human Nature* (Reading, MA: Addison-Wesley, 1988), pp. 63-98.

[39]Ervin Staub, *The Roots of Evil: On the Origins of Genocide and the Evolution of Caring Societies* (Princeton: Princeton University Press, 1989): pp. 274-283.

[40]Warren Bennis, *On Becoming a Leader* (Reading, MA: Addison-Wesley, 1988).

[41]Herbert Kelman and Lee Hamilton, *Crimes of Obedience* (New Haven: Yale University Press, 1989), p. 338.

[42]Ibid.

[43]Sarah A. Conn, "Protest and Thrive: The Relationship Between Global Responsibility and Personal Empowerment," *New England Journal of Public Policy* (Spring/Summer 1990): 163-178.

[44]Chellis Glendinning, "The Anatomy of Heroism for the Nuclear Age," *Bulletin of the Peace Studies Institute* (Manchester, IN: Manchester College, 1988).

[45]Mary M. Watkins, "In Dreams Begin Moral Responsibilities: Imagination and Action," *Facing Apocalypse* (Dallas: Spring Publications, 1986), pp.70-95.

[46]Doug McKenzie-Mohr and James A. Dyal, "Perceptions of Threat Collective Control: Tactical Efficacy and Competing Threats as Determinants of Pro-Disarmament Behavior," paper presented at the 96th annual convention of the American Psychological Association, Atlanta, GA (August 1988).

[47]Cristine Russell, "Risk vs Reality: How the Public Perceives Health Hazards," *Washington Post* (14 June 1988).

[48]Lisa Peattie, "Economic Conversion as a Set of Organizing Ideas," *Bulletin of Peace Proposals* 19, No. 1 (1988): 11-20.

[49]Ted Taylor, unpublished personal interview with Melissa Everett (1987).

[50]For more information contact Institute for Defense and Disarmament Studies, 2001 Beacon St., Brookline, MA 02146.

[51] Stanislav Grof (Ed.), *Consciousness Evolution and Human Survival* (Boston: Beacon Press, 1986), pp. 239-250.

52William Irwin Thompson, "Pacific Shift: The Philosophical and Political Movement from the Atlantic to the Pacific," in Grof, pp. 218-238.

4
Economic Conversion in Perspective: The Values and Ethics of Engineers

Eve E. Gagné

Abstract

Although we often assume that values determine our preferences and life choices, little empirical evidence supports this view. Rather our circumstances limit, and to some extent determine, the values that we hold. One important reason for this is that values are typically employed to justify choices and circumstances *a posteriori* or after the fact. Similarly, we find this to be true with engineers. Their values are to some extent determined by socioeconomic circumstance and by their personal and professional lives.

Economic conversion can be expected to create temporary unemployment problems for defense-industry engineers, especially for older workers with highly specialized technical skills. Employment opportunities will be available in the development of surveillance equipment, medical equipment, engineering software, and in joint military-commercial endeavors. Military engineers' adjustment problems will reflect the level of their personal resources for dealing with temporary unemployment, relocation, reduced pay, and seniority.

Military engineers will not have the resources available to resist economic conversion because they have effectively resisted unions, and their guiding professional organizations probably do not have sufficient political clout. In addition, the ever-increasing numbers of women and minority engineers will hold less conservative values than their predecessors, and be less interested in national security and more interested in social equality and global issues.

Engineers experience below-average job satisfaction, except when they are engineer-managers, which provides them with average job satisfaction. Since small companies are more likely to provide significant managerial responsibilities, job satisfaction may be somewhat higher in civilian firms than in defense. Greater involvement of engineers with end users could be expected to provide rewards for humanistic engineering projects that might outweigh the financial rewards of defense work.

Educational institutions should provide more broadly based undergraduate curricula, including more liberal arts, social, and behavioral sciences. Engineers of the future will need to work on interdisciplinary multicultural teams in which global marketing of products specially tailored to local conditions will be essential. The expectation that engineers will earn a master's degree, along with a bachelor's degree, may become appropriate.

The curricula that are appropriate for future engineers have not been studied. Longitudinal studies of engineers may be necessary to determine which programs are most desirable. For example, does an emphasis on interaction with end users affect career decisions? Empirical research will help to determine which, if any, re-employment services might be useful to displaced engineers.

Economic Conversion in Perspective: The Values and Ethics of Engineers

Values are determined in large part by circumstance. For example, gold loses its value on a sinking ship—not one passenger would exchange a life vest for a gold brick under such circumstances. In order to understand what engineers value, we must therefore understand their personal, professional and financial circumstances.

Values provide both a "blueprint for personal action,"[1] and an "element of predictability to social life."[2] Since World War II, one such value in the U.S. has been the importance we have placed on national defense. Many engineers have devoted their professional lives to nurturing a national defense. We and engineers choose and sustain values such as these within a particular context, which may be religious, social, or existential. When the context changes, as is now the case with economic conversion, significant, and often agonizing, shifts in priorities may be required of those so affected. Depending on how readily engineers recognize the change in values outside their immediate environment, their values may facilitate or interfere with their adjustment to personal and professional disruptions accompanying economic conversion. Values associated with globalism are now overtaking more parochial national values, and engineers who do not change their personal values accordingly may find themselves in poor health, mentally and financially.

A value is an enduring emotionally connected belief about a conduct or condition that is desirable[3] and that may lead to action.[4] Everyone holds values that are similar to different degrees;[5] thus the examination of values must always be comparative, examining how values are prioritized by particular individuals or groups of individuals. An understanding of engineers' values can be expected to accomplish the following: (1) uncover concerns for engineers facing economic conversion; (2) help predict engineers' likely reactions to economic conversion; (3) identify acceptable alternatives for engineers converting from the military to the civilian economy; and (4) identify social and educational programs that may meet engineers' needs.

A phrase that aptly describes the shift in values necessary for conversion comes from Isaiah: "... [nations] shall beat their swords into plowshares" (Isaiah 2:4). Stix[6] and Thee[7] both refer to the quote in their discussions of economic conversion. One of the values making swords necessary is that "... they worship the work of their own hands" (Isaiah 2:8). Making plowshares is possible only after "[t]he wolf also shall dwell with the lamb ... and a little child shall lead them" (Isaiah 11:6). The values of the little child stand in stark contrast to the modern preoccupation with technology and wealth. Although there are indications that economic conversion may be underway, it seems doubtful that this movement will last unless accompanied by a profound change in modern values. Such changes in values do not presently appear likely.

By examining engineers' values and ethics we can determine the possible effects of economic conversion. There are three value-related questions that need to be addressed. First, will engineers' conservative values make employment problems difficult to bear? Second, can engineers be helped to participate in economic conversion? Since professionalism is important to engineers, this orientation may help them adjust to economic conversion while permitting them to actively direct the practical changes required by conversion. Third, how will engineers impact the economic conversion process, given their ethics and values? Obviously, it would be desirable for engineers to constructively embrace a national policy that is presently gaining momentum. Research indicates that values can be redirected under certain circumstances. Engineers who have devoted their professional lives to the national defense may now be required to eschew such work in favor of the development and production of commercial products, thus requiring a shift in work values.

Economic conversion will likely change the lives of engineers on both a personal and professional level. At present the broader economic implications of economic conversion are better understood than the implications for individuals. In its initial stages at least, economic conversion will probably disrupt carefully crafted engineering careers with the added disadvantage of reducing earnings for those affected. Engineers who are displaced by economic conversion must be provided with psychological coping mechanisms, perhaps including discussing values. Simply providing engineers with career guidance or retraining will not address the primary issues for professionals who are totally committed to their work.

The Employment Outlook for Engineers: Conversion, Recession, and Valued Jobs

When little agreement about definitions of "engineer" exists and when data often do not distinguish between engineering specialties, scientists, and engineers, employment data on engineers are difficult to obtain.   For example, the U.S. Labor Department classifies experienced technicians as engineers; one-third of "engineers" in federal statistics do not hold an engineering degree.[8]   However, we can infer the number of engineers employed based on a variety of available statistics.

The defense purchases of machine tools was 3% of the market share in 1977 for all machines. This increased to 34% of market share in 1985. According to Henry and Oliver,[9] the employment allotted to military contracts can be estimated directly from market share of total industry output; thus, the same 34% of those employed in the machine tool industry could be expected to be working on defense contracts. Corroborating this, the industries with defense-related employment at or above the 10% level were all in durable manufacturing, which includes defense equipment, metals, and metalworking equipment.   Nearly one half of aeronautical engineers, and 19% of metallurgical engineers, were employed in defense production in 1985.[10]

Although NASA was expected to continue enjoying double-digit budget growth, in mid-May of 1991 the House appropriations subcommittee killed funding for the planned $40 billion Space Station Freedom, while also sharply reducing spending on the Superconducting Super Collider. In addition, SDI will be slashed from President Bush's requested $4.5 billion to only $2.7 billion.[11] Tens of thousands of engineers and scientists who would have been employed in "big science" projects will probably be seeking less lucrative jobs, with less advancement potential. Many of the engineers who will be affected are in high-demand jobs and can be expected to be more productive, though less well paid, in private industry. [12]

Employment in the defense industries declined by 2% in 1991,[13] and even the Persian Gulf War hasn't deterred defense budget cuts for 1992.[14]   The Defense Budget Project expects military spending to be reduced by 13.6% by 1995: one million defense-related jobs may be lost.[15] The bad news has been well summarized by economic forecaster DRI/McGraw-Hill, which expects 20% of jobs in defense-related industries to be lost by 1995, compared to peak 1987 levels. It now appears that employment in the defense   industries will decline by 2% in 1991.[16] Similarly, the Electronic Industries Association (EIA) expects a real dollar decline of 2% per year in the defense budget through fiscal 1994; electronics and material development should fare relatively well, while mechanical manufacturing and aeronautics will fare badly, though commercial aviation may offset this.[17]   The Institute of Electrical and Electronics Engineers

(IEEE) predicts that 55,000 of the 240,000 engineers in the defense industry could lose their jobs by 1995.

The aerospace industry is particularly hard hit by defense cuts. Aerospace jobs in both the defense and civilian sectors decreased by 2% for 1991.[18] In 1990 alone, 20,000 professionals lost their jobs and the total number of aerospace jobs declined from 1,328,000 in 1987 to 1,266,000.[19] Job losses are expected to continue through 1994. The present employment picture is not as bleak as it was in 1977, when the government accounted for 80% of U.S. aerospace sales. Today, government spending accounts for only 55% of sales.[20] An economic recession and fear of flying due to terrorism associated with the Gulf War concomitantly reduced demand in the commercial sector.

The major concern against economic conversion is that 726 federal laboratories, which employ more than one-sixth of the engineers and scientists in the United States, will disappear.[21] Currently, one-sixth of the total budget for federal labs is received by nine multiprogram labs in the Department of Energy. These labs include Sandia National Laboratories in Albuquerque, N.M., with a budget of $1.2 billion and employing 3601 researchers; Lawrence Livermore with a budget of $1.09 billion and employing 3304 researchers; and Oak Ridge National Laboratory, with a budget of $5 billion and employing 2073 researchers.[22] If extension facilities, hospital laboratories, and engineering test centers are included, federal labs cost $21 billion a year; otherwise, the cost is $15 billion a year.[23]

Although most engineers will not be directly affected by conversion, engineers most affected will be those with highly specialized technical experience in the defense industry. Many of these engineers will be older individuals who will find a job loss particularly difficult.

Twelve states account for 64% of the nation's defense jobs Since past attempts at diversification by defense industries have failed, most displaced engineers will need to seek new employers, which may involve moving.[24] On the positive side, the high geographic density of employment problems will provide engineers with easy access to support groups.

A book appropriately entitled *The Great Job Shake-Out* examines the employment possibilities for the 1990s and finds that the one type of business that tolerates a high frequency of startups, has a good survival rate, and an excellent chance of economic growth is engineering and architectural services.[25] Despite the difficulties engineers will endure during economic conversion, this is still a career path that enjoys one of the lowest unemployment rates, even in difficult economic times.

A summary of the data on economic conversion and engineer employment indicates the following: (1) many experienced engineers will be displaced from their current positions in the defense industry, because of both conversion and recession; (2) unemployment among engineers will continue to be about one-third of the unemployment rate in the general population; (3) the vast majority of engineers will be re-employed within less than a year; (4) re-employed engineers will lose seniority privileges; (5) most re-employed engineers will work in small civilian firms (fewer than 1000 employees); and (6) most re-employed engineers will earn significantly reduced salaries. Table 1 summarizes economic and social values underlying the problems of economic conversion as well as possible reorientations or solutions.

Table 1

Economic Conversion: Problems in Economic and Social Values

PROBLEMS	VALUES/REASONS	POSSIBLE SOLUTIONS
National security emphasis.	Centralized decision-making, secrecy, nationalism.	Economic security, global emphasis, technology transfer.
Reduced funding of federal labs.	Trade deficit, smaller budget, arms control.	Joint ventures between labs and industry encouraged by Congress.[26]
		Conferences now planned to bring industry in contact with lab researchers.
Short-term goals in research and development.	Cost of capital, antitrust rules applies to U.S. only.	Product development in areas of lower profitability and social need.
Slow technology transfer. (U.S. companies get 55% of electronics patents in 1987 compared to 80% prior to 1984.)	Emphasis on military and defense with poor transfer potential to commercial markets.	International cooperation in commercial ventures.
		Federal labs create centers for entrepreneurs (140 startup companies commercialized DOE technology from 1985 to 1990).

Unemployed Engineers

In the early 1970s, 80,000 electrical and aeronautical engineers were assisted in finding new employment through the joint efforts of IEEE and the American Institute of Aeronautics and Astronautics (AIAA). Working with the U.S. Department of Labor, out-of-work engineers conducted seminars on positive mental health and the basics of job hunting for their colleagues; many companies also provided such services to laid-off employees. Unfortunately, efficacy studies of these interventions do not appear to have been conducted. It would be particularly useful, given the current situation, to know what works—and doesn't work—when attempting to help engineers redirect their careers.

In 1989, 1.4% of U.S. engineers were unemployed, but by the end of 1990 that number nearly doubled to 2.2% of the nation's 1.9 million engineers being unemployed, and the numbers are still increasing.[27] Although big companies dealing in mainframe computers and aerospace products have experienced a large number of layoffs, small companies, particularly those in the medical and computer software fields that employ fewer than 1000 people, are growing rapidly.[28]

Engineers who lose their jobs will probably be hired by small companies at significantly reduced salaries, but they will be re-employed. The economic recession is responsible for more engineering job losses than are defense cuts.[29] In fact, "only one out of seven engineers—one out of six EEs—is supported by defense," according to Robert Rivers of the IEEE Manpower Committee.[30]

Japan productively employs a far greater number of engineers, per capita, than the U.S.; one conclusion might be that engineer unemployment in this country is probably due to poor planning, management, and distribution of the work force. Given the rapid technological, social, and economic changes of the information age, schools of engineering must prepare future engineers for career changes. Such programs must also train engineers in job search skills so that midcareer moves will not be overwhelming for them.

The new global economy can be expected to provide highly skilled engineers with more employment opportunities. But the internationalization of professional work will also result in more competition, often from engineers in other countries, and salaries for most engineers may be reduced as a result. The future of engineering professionals can be expected to be vastly different from the past, but little attention is being given to the changes in professional education that such developments require.

Values, Stress, and Unemployment: Implications for Engineers

Most engineers stay with an employer for less than four years on average.[31] But planned and voluntary job changes made for advancement purposes are obviously very different from the economic conversion layoffs of the 1990s when engineers will often be required to accept reduced pay and rank in order to be re-employed.

Job loss among engineers, as with most U.S. citizens, is seen as devastating; on standardized measures of psychological stress, it ranks among the most stressful of life events.[32] Unfortunately, when the paid work career is lost or must be changed engineers often see themselves as losing their greatest investment.[33] This need not be the case, however. At least four careers can be identified—including not only paid work, but also family work, avocational work (or hobby), and community work. A career involves the development of talents and abilities; it is a series of productive endeavors in the service of others.[34]

Although most engineers are husbands and fathers,[35] the only available figures on the matter[36] indicate that only 47% of engineers believe family relationships are more important than their paid work career, but 58% of engineers believe that most engineers think their professional career is more important. On the positive side, Zussman[37] found that engineers are nearly twice as likely as nonengineers to participate in family-based groups such as school service organizations (18%, compared to 10%) and significantly more likely to volunteer for youth groups (11% compared to 8%). Zussman's findings are difficult to explain in light of the Perrucci and Gerstl research. One possibility is that family life has become more important in the U.S. since the Perrucci and Gerstl study. This is supported by recent findings. In the 1980s adults 30 years old and younger were dramatically less interested in self-fulfillment than same aged individuals a decade earlier (5.5% in 1986 versus 18.7% in 1976) and more concerned with self-respect (18.9% vs. 14%), warm relationships (26% vs. 19.9%), and fun-enjoyment-excitement (15.7.% vs. 7.8%).[38]

There are indications that the pendulum is about to swing in the other direction: a 1989 Gallup poll of college students found a majority agreeing with "making a lot of money is my most important goal"; in comparison, Gallup polls of college students in the 1960s indicated that developing a "satisfying philosophy of life" was considered the most important goal. It is of course possible to justify engineers' often placing greater emphasis on paid work than on family relationships—for example, men are often socialized to believe that they have primary responsibility as breadwinner for their families. In the examination of values, although assigning "rightness" or "wrongness" to values is inappropriate, examining functional versus dysfunctional values is important.

For engineers who face major employment disruptions, including demotions, reduced pay, and periodic unemployment, an emphasis on paid work as the primary career is dysfunctional not only for engineers but for their families and communities. Adding to this problem the intense and highly specialized education required of engineers in college does not expand their general interests. In fact, engineering students are less likely to engage in extracurricular activities than the average college student.[39] Obviously, engineers who have invested time and effort in cultivating family relationships, community ties, and avocational interests are better prepared to cope with the stress of unemployment. Of particular importance is the social support of the family—the stress of unemployment may destroy an already weak family structure—thus leaving the unemployed engineer with yet more stress.

The socialization process that produces the professional engineer reinforces tendencies toward narrow interests. One of the best studies of professional socialization found that physicians are socialized to focus on "surviving" by being placed in residency programs that put their personal or professional lives out of control.[40] Since they are given extremely large patient loads, residents learn to value efficient dispatching of patients.[41] Residency programs change the values of physicians-in-training, and early professional experiences appear to change the values of engineers. The traditional "sink or swim" model that engineers face in industry implies that they are not personally important and reinforces technical values at the expense of interpersonal values.[42] The engineer quickly learns that his contribution consists of a narrow set of tasks, and salary compression announces the obsolescence of his skills.[43] Self-valuations as specialized technical workers with few interests could be expected to thwart engineers' employment-seeking behaviors when faced with economic conversion layoffs.

Research with law students indicates that the values learned in law school are changed by the marketplace.[44] Economic conversion can be expected to change the values of engineers as well, but particularly the values of younger engineers who have no investment in defense work.

In studying job loss among engineers, technical managers, and scientists, Jacobson hypothesized one of two stress models might help them to readjust.[45] Engineers' adjustment to job loss might be primarily dependent on their resources (transactional model), or primarily dependent on changes in their beliefs and values. Jacobson found that job loss may have varying meanings and cannot be understood out of context. One factor is a technical worker's resources, or lack of resources, which largely determine the extent of anxiety and dysfunctional emotional states that are experienced. When technical workers felt identity confusion or questioned the meaning of work, these problems stemmed from work experiences rather than from job loss. Jacobson's findings are in agreement with Rokeach,[46] who has observed that behavior is motivated both by the value of the object (a job that is lost) and by the interpretation of the situation (now we will lose our house and have no place to live). Engineers with considerable savings and few debts would react to job loss differently from engineers with considerable debts and few savings. Therefore, a focus on helping engineers change their beliefs and values will help them adjust to economic conversion only when they perceive they have adequate financial and social resources.

Engineers have the poorest severance package of any professional group, an average of 5.4 months. That they take about as long as other groups to find a new job, with an average of 6.3 months, could be another factor that increases levels of stress and anxiety when engineers lose their jobs.[47] Job loss affects the unemployed worker's family, including relationships with the spouse and children.[48] The severity of the stress experienced with unemployment can even be expressed in child abuse: families experiencing chronic unemployment are more likely to abuse their children.[49] In order to reduce the severe stress that unemployed engineers experience, advance warning of plant closings[50], outplacement counseling, and family-crisis intervention can help.[51]

Unemployed Engineers' Values

According to a 1990 study conducted by an executive outplacement firm, Drake Beam Morin, engineers require an average of 6.3 months to locate a new job, while older engineers require more time. Job openings for engineers remained steady at 4%, but when engineers obtained new jobs, their average salary was $65,550, compared to an average of $68,619 in their old jobs.[52]

Engineers have often neglected to network with professionals outside of their companies and to expand their skills from narrow specialties, with an eye on the job market; thus they are often ill-prepared to find new employment.[53] Some professional organizations have organized resume data bases that are available to potential employers. The American Engineers for Social Responsibility, for example, wants to help engineers convert their skills from defense to civilian work.

Theory and research in values acquisition supports that values freely chosen from among alternatives, publicly affirmed and acted upon, are those that are most important to the individual.[54] In keeping with this, Pfeffer[55] found that engineers are more likely to be committed to employment positions that they have held for some time and to be most satisfied when their job was difficult to obtain despite numerous other offers. Dislocations caused by economic conversion can be expected to lower engineers' morale, even if comparable jobs are obtained. Engineers are known to have particularly high needs for order, endurance, and achievement.[56] Layoffs and the resulting interrupted careers can be expected to be particularly disturbing to them.

Engineers' conservative values and their personal identification with corporate values have made unionization rare.[57] Thus engineers who lose their jobs during the period of economic conversion will not receive whatever benefits unions can provide to threatened workers.

The Promise Of Conversion: New Global Opportunities

Many engineers whose jobs have been impacted by economic conversion and/or recession are retooling for software engineering, which pays up to three times more than other engineers' average salaries.[58] There is also currently a shortage of engineering professors. Defense engineers with advanced degrees could probably find employment in colleges and universities, though at reduced pay.[59] Many civilian engineers who have been working on military projects can expect to work on technologically advanced surveillance equipment to ensure national security in the face of military threats from other countries.[60]

A factor that may serve to blunt the effects of economic conversion on older engineers is that a majority of engineers in the U.S. are managers by age 65 and 73% of engineers have significant managerial duties by age 50.[61] Managerial positions are probably more interchangeable between the civilian and military sectors than more specialized, technical positions. And for engineers who make a point of obtaining licenses, consulting appears to be a promising venture, especially in the South, according to a 1991 survey by the American Consulting Engineers Council. Another growing option for engineers is work in other countries. According to the Accreditation Board of Engineering and Technology (ABET), cross-licensing is expected to make engineering a truly international profession.[62] It does appear that under some circumstances, such as immigration to Israel, engineers are able to adjust emotionally to job loss by lowering their expectations, committing themselves to work, and concentrating on short-term goals.[63]

A Shortage of Engineers?

Despite current concerns about engineer unemployment, a number of forecasters, including the U.S. Bureau of Labor Statistics, forecast a shortfall of 300,000 to 700,000 engineers within the next decade.[64] Expectations of a shortage are based on a moderate growth rate for the economy and the fact that although undergraduate degrees in engineering have remained at 4% of all undergraduate degrees awarded for the past 30 years, the "baby bust" years are expected to reduce the numbers of undergraduate students. About 66,000 engineering bachelor's degrees were granted in the U.S. in 1990. Such statistics do not take into account a possibly dramatic increase in nontraditional students, including older students, women, and minorities.

Although programs are currently under way to improve math and science performance among high school graduates so that they may successfully major in engineering, these efforts may be too little and too late.[65]   In Japan, 40 out of every 1000 college students choose an engineering major, whereas in the U.S., only 7 choose engineering.[66] Thus, the freeing up of engineers from the defense industry can be expected to provide sorely needed talent to the civilian sector.   Another reason that engineer unemployment should not become a major problem in the U.S. is that the number of undergraduate engineering degrees has been declining since 1983 except for a slight increase in 1988, reflecting the low birth rates in the 1960s and 1970s.[67]

In the near future, engineering education will face a crisis because one-third of engineering professors will reach retirement age in 1995.[68] Engineering schools already place heavy reliance on foreign nationals, who receive an ever growing number of U.S. Ph.D.s in engineering, which currently rests at 50% of engineering Ph.D.s, but who often have limited English-speaking skills.[69]

Economic conversion will create a temptation to use laid-off engineers as college professors, but this may be misguided.   The past experience of an engineer may have little relevance to future technology, and some engineers may be overspecialized for most college programs.[70] Rather than seeking technical experts, colleges and universities should probably seek engineers who are experienced in initiating and managing change.[71]

According to a study by the Office of Technology Assessment (OTA), the federal government has extensive influence on science and engineering programs since it funds most higher-education R&D in these fields and because it is a primary source of support for many graduate students, which amounted to 45% of engineering graduate students in 1967 and 20% in 1985.[72]   The OTA study notes that the numbers of Ph.D.s are directly related to the financial support available and the overall outlook for research funding.   A recent shift from federal fellowships (which foreign students may not receive) to research assistantships may have increased the accessibility of advanced graduate study to foreign students.

If a tight federal budget further restricts support of graduate students, further shortages may develop.   On the other hand, unemployed engineers from the military sector might be more likely to use personal resources to obtain a Ph.D. than engineers who are employed.   The recent dramatic increase of Ph.D.s in ceramic engineering obviously reflects market conditions.   Engineers may attempt to protect themselves against job loss by obtaining advanced degrees.   For example, a 17% increase in master's degrees in 1990 compared to 1989 may reflect this trend.[73]

Some engineering jobs are more attractive than others, which may compound shortage problems.   For example, manufacturing engineering is seen as a dead-end job, while product design is viewed as better paying, more glamorous, and more likely to provide advancement;[74] inequities are amplified when simultaneous engineering teams require manufacturing engineers (MEs) to work with design engineers.

Further complicating the picture of future needs for engineers is that the National Science Foundation found that only 66% of the high school class of 1980 who obtained degrees in engineering or in the natural sciences are now full-time employees and only about one-half of that 66% are employed in readily identifiable science and engineering professions.[75]

Engineers' Professional Values: Military vs. Commercial Sectors

Military components are typically overdesigned to ensure their ability to withstand extreme battlefield conditions.[76] Military equipment must first go through a development phase which only the military can evaluate and a production phase for which Congress allocates funding, permitting the purchase of a specified number of units. It has been suggested that a prespecified fixed budget prior to the development phase might result in a greater number of less sophisticated weapons, thus reducing the tendency to overdesign.[77]

The military emphasis on performance characteristics, with equipment often built to prespecified military requirements opposes the commercial market emphasis on operating costs.[78] As a result, dual-use technologies, or technologies with both commercial and military applications, are rare, with only 1% of the Navy's patents receiving commercial licenses in contrast to the Department of Agriculture, which owns 13% of patents.[79] Even when military technology has civilian applications, it tends to remain classified information for some time.[80] Both the defense and commercial markets now have a global focus.[81] As a result, most countries that desire sophisticated weapons may acquire them. Given the impact of economic conversion on their bottom lines, U.S. defense industries may be more determined than ever to export arms.

The New Melting Pot: Military and Commercial Mix

Successful foreign competition has made the federal government increasingly aware of its responsibilities to industry, and military services are being recruited for this new crusade. For example, the U.S. Navy's Best Manufacturing Practices Program (BMP) is promoting the best practices of commercial firms by reducing their production and shipping costs, thereby strengthening U.S. industry while reducing defense costs.[82] The new emphasis in industry on systems engineering, which reduces engineering changes and emphasizes ease of manufacture, has significantly reduced product cost and improved product quality. This and future efforts require a shared data base in which computer-aided design (CAD) becomes more than a design tool.

For its part, the Department of Defense (DoD) has developed data base management that allows engineers to evaluate the effects of product design on product users while providing information on the training needs this entails. Along similar lines, the Defense Department's Advanced Research Projects Agency (DARPA), comprised of 75 scientists and engineers with a budget of about $1 billion, has functioned as a venture capitalist for the DoD by funding high-risk commercial technologies with great military potential.[83]

Reduced military budgets mean that the DoD is no longer determining the nation's research and development agenda.[84] In coming years, military engineers will be required to adapt off-the-shelf commercial products to military use.[85] Engineers who work on military products will be encouraged to show greater interest in commercial applications as the federal government continues to foster international competitiveness for U.S. products. Japan and Europe have long enjoyed economic incentives for commercial applications from their governments. Because of the 1986 Federal Technology Transfer Act, federal researchers get at least 15% of the royalties from their inventions, and royalties from federal labs more than tripled in the first three years the law was in effect.[86]

Values: Coping Tools for Engineers

According to Rokeach, "All of a person's attitudes can be conceived as being value-expressive, and all of a person's values are conceived to maintain and enhance the master sentiment of self-regard—by helping a person adjust to his society, defend his ego against threat, and test reality."[87] Rokeach goes on to say that, "... when a person tells us about his values, he is surely also telling us about his needs."[88]

Psychologists have long assumed that basic values, such as physiological and safety needs, must be satisfied before attending to higher-level needs or values, such as self-actualization.[89] This has been surprisingly difficult to verify empirically, given the many levels of values. For example, changing social conditions, including new technologies, alter the conditions of choice and result in value change.[90] In 1976 the two most important values for women were security and self-respect, but as more women worked outside the home, their most important value shifted by 1986 to warm relationships with others.[91] The same study found that men between the ages of 40 and 60 placed less emphasis on security and a sense of belonging than same-age men in 1976 and more emphasis on a sense of accomplishment and warm relationships with others. This development may be promising for engineers caught in the throes of economic conversion. Engineers who value warm relationships with others may experience less stress when faced with employment insecurity. Engineers, more often than other professionals, come from the Midwest and from upper-lower-class families.[92] Parental values have been shown to be strongly related to social class, with middle-class parents emphasizing self-direction and working-class parents emphasizing conformity.[93] Thus, it is not surprising that engineers' conformity to corporate expectations is a recurring theme in the study of their values.[94] When disciplining children, middle-class parents are more likely to examine a child's motives and feelings, while low socioeconomic status (SES) parents are likely to focus on what the child did.[95] As might be expected, engineers prefer to work with facts rather than with feelings and people.[96] Most engineers in the U.S. are white males who find themselves working in the corporate world. These conditions and values tend to make engineers conservative.[97] Even as college students, engineers tend to be conservatives.[98]

Conservatives place greater emphasis on freedom and less value on equality than liberals or other groups[99] and this is also true of engineering students.[100] Persons who rank equality of high importance are likely to explain social problems, such as poverty, on the basis of structural problems in the social fabric.[101] Engineers, on the other hand, place great value on competence, especially technical competence, and are probably more likely to attribute problems in social stratification to personal incompetence.[102] Engineers' emphasis on freedom and de-emphasis on equality could be expected to result in a lack of concern for the "have-nots" and a concern for national security.[103] The implications of engineers' conservatism for economic conversion are obvious—national security, technocracy, and issues affecting the privileged are seen as more meritorious than issues affecting those who are impoverished and presumably incompetent. Since these attitudes are prevalent even among freshman engineering students, neither educational experience nor work experience produces this value system.[104] However, more exposure to the liberal arts and to the social or behavioral sciences might help engineering students become more empathic and concerned with issues affecting the "have-nots," including a greater concern for global issues.

Male engineers have more conservative values and attitudes than female engineers, especially when judging the desirability of having females in leadership positions.[105] Some evidence suggests that men in male-dominated occupations may "heighten the boundaries" between themselves and women by overemphasizing certain aspects of male culture, such as drinking and swearing, which can lead to the well-rationalized ostracism of women.[106]

Research comparing the values that attract men and women to engineering indicate minimal sex differences, yet very few women are in engineering programs.[107] Data indicate that women are not found in engineering because they experience too much personal and professional isolation in engineering education programs.[108] Black women may have a cultural background that gives them greater tolerance to that isolation since one-third of B.S. engineering degrees awarded to blacks in 1990 were awarded to women. In recent years, women have constituted about 15.4% of engineering graduates at the B.S. level, but doctoral degrees awarded have jumped from 6.8% in 1988 to 9.1% in 1990, and actual numbers from 88 in 1980 to 495 in 1990.[109]

Although investigators have noted that engineering is dominated by conservative professional organizations, some dramatic changes may be on the horizon.[110] Anecdotal evidence indicates that the demand for engineers who are women or minorities may be higher in 1991 than in the past eight years combined.[111] This influx of women and minorities can be expected to have a "liberalizing" effect on the traditionally conservative engineering profession.

Engineers tend to be pragmatic[112] and focused on their individual career goals, which tends to make them very mobile, often remaining with a company for less than four years, on average.[113] Frequent moves affect the personal lives of engineers by increasing their reliance on engineer peer groups for friendship and decreasing their participation in community affairs when compared to engineers who are not mobile.[114] An increased reliance on engineer colleagues as a primary source of friends reduces the importance of outside influences and probably has a compounding effect on values that seem to characterize the profession, such as conservatism.

Engineers' Attitudes Toward Work

Attitudes are far more numerous than values, but like values they are both cognitive and affective.[115] Attitudes are probably a function of basic values.[116] Since attitudes are more specific than values, attitudinal information is especially useful in examining the interplay between the engineering profession and economic conversion.

Unlike the more traditional professions, engineers hold a pro-business stance that de-emphasizes the importance of autonomy.[117] However, Pfeffer[118] found that autonomy was a consistently important predictor of engineers' attitudes toward their jobs, but experienced engineers are less optimistic about their ability to manage their own careers than are engineering students.[119] In fact, engineering students show a greater desire for autonomy than working engineers, who may accept the reality of low autonomy given organizational hierarchies.[120]

Engineers who are most satisfied with their work tend to have strong investigative and creative interests.[121] With experience, engineers come to see the extrinsic rewards of technical work as less appealing than originally.[122] The aspirations of working engineers better reflect their values and beliefs than do those of engineering students, but students who aspire to management seem to hold values and beliefs that predispose them to favor this over technical work.[123] Experienced engineers often aspire to occupations outside of the corporate world, such as consulting, academics, or owning their own business.

The lack of autonomy engineers experience in their usual corporate employment settings, along with the bureaucratic assignment of narrow and repetitive tasks, may help to explain why they perform below potential productivity[124] and experience less than average job satisfaction.[125] Until recently, there was reason to believe that the increased need for affiliation that older men experience,[126] at least in part, accounted for the move from technical to managerial positions, but more recent work found engineering students more likely to aspire to management than working engineers.[127] Not surprisingly, then, engineer-managers are more likely to be satisfied with their jobs than are other engineers, but their level of job satisfaction is only average compared to men in general.[128]

Engineers with high achievement, affiliation, dominance, and status needs are especially satisfied with the engineer-manager role.[129] Using a values rating scale that was based on Schein's[130] concept of "career anchors," Rynes, Tolbert, and Strausser[131] found that engineering students and alumni who preferred the *managerial anchor* liked being involved in policy making, getting promotions, and earning more money than others with a similar education. Those with a *technical anchor* were interested in getting patents, making breakthroughs, publishing, becoming authorities in their field, and doing the work they preferred.

Certainly, other important incentives to move from the technical to managerial positions are managers' higher salaries, greater opportunities for advancement, and added power and autonomy to execute responsibilities, even in companies with dual career paths.[132] Dual career ladders have been described as a desperate attempt to retain technical experts by redefining failure to enter managerial ranks as success.[133] Engineers who experience layoffs as a result of economic conversion will likely find increased managerial responsibilities. The increased sophistication of production, emphasizing concurrent engineering and flexible manufacturing systems, means that few engineers will work in isolation and more engineers than ever will enjoy interdisciplinary and managerial responsibilities.

Since management selects for promotion to engineer-manager those engineers who are "loyal" to the company[134] and conventional in their interests,[135] management tends to reproduce itself. Engineers exhibit corporate values to a greater extent than scientists[136] or production and clerical workers,[137] and they are more likely to be promoted to high-level corporate positions, despite their lack of training and experience in "people skills" and other managerial responsibilities. The fact that engineers have "thing" or "data" training rather than "people" training can create problems in the switch to a managerial position.[138]

As in the general population, reality-based experiences shape the values of engineers.[139] Thus, engineers who have been employed in defense industries have valued that work, but they could be expected to value work for civilian projects, given that opportunity. In a study that investigated this, the researcher could not identify the factors affecting attitude change among engineers.[140] Surprisingly, both company subunits that spent more time together and units that did not socialize outside work shaped job attitudes. Pfeffer hypothesized that isolation from workers in other units and/or the need to conform with co-workers in the face of uncertainty might account for work unit effects on job attitudes. This uniformity of attitudes among unit workers could be expected to facilitate employment intervention programs with engineering units that are being dismantled due to economic conversion, recession, or changing technological needs.

Engineering Ethics

Ethics are moral values—values that address the issue of right and wrong. The exercise of engineering ethics is most visible in whistleblowing activities.[141] Whistleblowing requires great conviction and fortitude since whistleblowers often pay a high price in career terms for their activities. This said, however, whistleblowing is reactive rather than proactive and usually comes after the fact, involving blame rather than positive planning. That whistleblowing should figure so prominently in the engineering profession is perhaps the result of engineers' belief that "the engineering profession is composed of individual engineers, so that engineering will be responsible if its individual member engineers are responsible."[142]

Whistleblowing can be costly to individuals and to society when existing problems are not anticipated, or when individuals are reluctant to blow the whistle post facto. A surgeon would not operate before being fully informed of the patient's condition. Professional ethics should require similar assurances before engineers participate in projects. Engineers who make a serious effort to understand both the technical and social implications of a project might be able to avoid participation in projects with the potential for great harm.

We create our value systems by selecting specific questions out of a universe of possible questions. For example, only religious answers are possible when given religious questions and only empirical answers are possible given empirical questions.[143] The paradigm that is selected "locks in" the user. Paradigms are not self-correcting but rather self-perpetuating. Self-perpetuation of paradigms is particularly insidious when the sociocultural context is homogenous, "... because he is working only for an audience of colleagues, an audience that shares his own values and beliefs, the scientist can take a single set of standards for granted."[144] According to Raskin and Bernstein:[145]

> ...the most dangerous partial truth about science and technology is belief in their neutrality. This is especially obvious in technology, for the act of making things is laden socially with import.

According to Zussman,[146] engineers are significantly more likely to participate frequently in church groups than average (30% of engineers compared to 20% of a national sample). A study of religious values found that "[f]our of the major Protestant groups (Episcopalians, Lutherans, Presbyterians, and Methodists) rate patriotism ahead of values concerning salvation, helping others, or achievement."[147] Religious groups that rate patriotism as more important than salvation or helping others would probably provide strong support for those employed in the defense industry.

Changing Engineers' Values

Although engineers' natural pragmatism is likely to facilitate their adjustment from a defense economy to a peace economy, changing their conservative corporate values now might spare them the pain of reacting negatively to layoffs while at the same time providing them with skills that will help them to find new work. Values are very difficult to change and the associated behaviors are even more resistant to change.[148]

The "value confrontation approach" has shown to have long-lasting effects about 59% of the time.[149] The value-confrontation approach to behavior change[150] typically involves the arousal of self-dissatisfaction by causing subjects to become aware of the discrepancy between their professed values and their actual behavior when compared to others whose actions reflect their values. For example, Sawa and Sawa[151] obtained students' self-reports by asking them about the time spent exercising compared to the importance they placed on health and the amount of exercise they felt they *should* have. Students were given a lecture implying that hypocrisy might lead to unhappiness when beliefs and goals were not congruent with actions. The value-confrontation was effective when subjects placed great value on health, found their actual behavior inconsistent with their ideal behavior, and were dissatisfied with their current behavior. Such an approach might be effective when examining engineers' ethics and behaviors in educational settings.

Another approach to values change has been the "values clarification" approach.[152] This requires that individuals reflect on their values, choose a value from among alternatives, prize the value, publicly affirm it, act upon it, and make it an important part of their lives.[153] Thus, if an engineer values the professional career over the family career, the implications of such a choice should become clear after sufficient reflection, particularly when such reflection occurs in a group context. This approach would be appropriate content for undergraduate programs.

Because of the nature of their work, engineers may be in greater need of various types of values education than most professionals. Heiddeggere[154] noted that in modern life everything is functioning and "technicity increasingly dislodges man and uproots him [from place and social tradition]."[155] The result is a split between fact and value, which creates an existential dilemma:

> [Man] makes himself such that he is *waited for* by all the tasks placed along his way. Objects are mute demands, and he is nothing himself but the passive obedience to these demands.

> But ... existential psychoanalysis ...must reveal [to man] that he is *the being by whom values exist*. It is then that his freedom will become conscious of itself and will reveal itself in anguish as the unique by which the world exists.[156]

One problem with values seems to be that many people, including engineers, appear to be overwhelmed by the complexity of values in a technological world, where man creates himself through his technology, e.g., as a being who does or does not reproduce, as a being who does or does not fly, etc. Numerous authors have documented the difficulties U.S. citizens are experiencing in trying to establish satisfying values in a world where this is the case.[157] The current trend seems to be ethical minimalism[158] or value relativism.[159]

Basically, minimalism holds that ethical values are a matter of personal preference, each person being entitled to decide rightness and wrongness based on personal feelings, the only rule being that no one may infringe on another's personal preference.[160] Roubiczek[161] criticizes minimalism by pointing out that relative values, which must be defined according to the larger social purpose, are confused with personal values, which are purely subjective. The resulting problem is well illustrated by a recent heated exchange between a talk show host in Chicago and a guest on that show, who informed the interviewer that he didn't agree with the U.S. Constitution. This minimalist guest saw his personal feelings as sufficient grounds for ignoring Constitutional rights.

Engineering Education: As It Is and As It Should Be

The current engineering science core appears to have evolved from the Grinter study[162] and may now have outlived its usefulness. As technology and global competition continue to heat up and expand, new educational tools and programs are required. Presently, the Accreditation Board for Engineering and Technology (ABET) works with the professional organizations that support it to develop appropriate curricula for engineering specialties.[163] This process probably ensures very conservative changes in engineering education. Alumni must be regularly and frequently tapped for advice on program improvements, and it would seem prudent to include advice and suggestions from diverse cultural groups and from other academic disciplines, in a world with an increasingly interdisciplinary and global focus.

Science and engineering coursework accounts for more than three-quarters of undergraduate engineering requirements while social sciences and humanities account for fewer than one-fifth of courses taken.[164] In his summary of survey findings, Zussman also notes that most of the theoretical knowledge that engineers learn in college is not used in practice. For example, engineers report "never" using information from required coursework in calculus, differential equations, and complex variables.[165] Thus, there appears to be room in the undergraduate engineering curriculum for a more broadly based educational program that might be more useful, not only to engineering students but to the consumer base they will ultimately serve.

Consumer psychologists have long recognized the importance of values derived from our vision of the "good life" in determining consumer buying preferences. In fact, most of the research on values is now being conducted by consumer psychologists. The laddering method,[166] uncovers an empirically based network of personal values. This type of study, commonly used for advertising purposes, would help educational institutions understand interrelationships between students' values and college curricula. This would clarify certain potential problems. For example, do students select engineering in order to avoid dealing with ideologies different from their own conservative outlook? Does the current curriculum encourage the maintenance of narrow interests, perhaps making it difficult for engineers to be effective in their technical and managerial roles?

If the engineering curriculum fails to expand its focus, engineers, who have been shown by the empirical research to value managerial positions, will be left out of an important new role, that of global manager.[167] The new global economy will require managers who have multicultural and cross-disciplinary skills. What may be called the "third industrial revolution" has produced technology that requires a local focus on a global scale.[168]

Values in Engineering Education

Whereas engineering programs have traditionally prepared students for highly specialized employment in a corporate setting, the future of engineering will depend on multifunctional teams. Multifunctional teams will include not only engineers from a variety of specialties but business managers. Educational programs should reflect these changes.[169] Teams of the future are likely to be multicultural, and working on such teams will require that students understand and value cultures and points of view other than their own.

The social sciences and the liberal arts would surely help engineers gain the skills needed to work in a multicultural environment. The currently available empirical research indicates that engineering students generally are white males from upper-lower-class families who have very limited cultural and social experiences. This points to the need for a broadly based educational program that is vastly different from the current undergraduate science and math curriculum. As the engineering skills required of practitioners grow in depth and breadth, perhaps a master's degree should be required of practicing engineers, with the graduate degree providing highly specialized skills and the undergraduate degree providing broadly based curricula. Knepler[170] suggests that the liberal arts should encourage engineers to see themselves as instigators—as well as instruments—of change.

Currently, engineering students are often required to design projects that are evaluated on technical merit, with growing concern that such projects be evaluated on business and life-cycle issues, including cost and manufacturability.[171] In a social science and humanities environment, attempts might be made to encourage the engineer to become involved with end users. Engineering projects might also be evaluated in terms of social usefulness. For example, a project that develops a mechanical bartender that mixes alcoholic drinks might be graded less favorably than a robot that can fix meals for quadriplegics. Encouraging engineering students to follow their projects to the end-user point might encourage them to develop products that are somewhat lower in profitability but higher in socially redeeming qualities, substituting engineers' lesser monetary reward with greater personal satisfaction. Such a development might improve engineers' satisfaction with their professional work, which is well below the average for other U.S. professionals. As an alternative to a large number of social science and liberal arts courses, topics such as engineering and society and engineering ethics could be incorporated into existing courses.[172]

Notes

1 I. Yalom, *Existential Psychotherapy* (New York: Basic Books, Inc., 1980).

2C. Kluckholm, "Values and Value-Orientation in the Theory of Action," in T. Parsons and E. Shils (Eds.), *Toward a General Theory of Action* (Cambridge, MA: Harvard University Press, 1951).

3M. Rokeach, *Understanding Human Values* (New York: Free Press, 1973).

4N. Feather, *Values in Education and Society* (New York: Free Press, 1975).

5Rokeach, *Understanding Human Values.*

6G. Stix, "From Swords to Plowshares," *IEEE Spectrum* (November 1989): 45-49.

7M. Thee, "Swords into Plowshares: The Quest for Peace and Human Development," in L. Dumas and M. Thee (Eds.), *Making Peace Possible: The Promise of Economic Conversion* (New York: Pergamon Press, 1989).

8L. Guy, "AES - If Not Now, When?" *Engineering Education* (January 1991): 32.

9D. Henry and R. Oliver, "The Defense Buildup, 1977-1985: Effects on Production and Employment," *Monthly Labor Review* (August 1987): 3-11.

10Ibid.

11A. Pasztor, "Gulf War Win Isn't Furthering Arms Spending," *Wall Street Journal* (20 May 1991): C16.

12R. Blank and E. Rothschild, "The Effect of United States Defense Spending on Employment and Output," *International Labour Review* 124 (1985): 677-697.

13E. Phillips, "Commercial Aerospace Companies Limiting Employment of Engineers," *Aviation Week and Space Technology* 134, No. 9 (1991): 61-62.

14Pasztor.

15J. Ellis, E. Shine, D. Griffiths, B. Carlson, K. Pennar, and E. Ehrlich, "Who Pays for Peace," *Business Week* (2 July 1990): 65-70.

16Phillips, "Commercial Aerospace Companies Limiting Employment of Engineers."

17G. Kaplan, "Unanswered Questions," *IEEE Spectrum* 26, No. 11 (1989): 70.

18Phillips, "Commercial Aerospace Companies Limiting Employment of Engineers."

19E. Phillips, J. McKenna, S. Kandebo, and D. Quast, "Recession, Military Reductions Force U.S. Aerospace Firms to Cut Payrolls," *Aviation Week and Space Technology* (4 March 1991): 52-55.

[20]T. Bell, "90s Employment: Some Bad News, but Some Good," *IEEE Spectrum* (December 1990): 32-43.

[21]J. Adam, "Federal Laboratories Meet the Marketplace," *IEEE Spectrum* (24 March 1990): 39-45.

[22]Ibid., 40-41.

[23]Ibid., 39.

[24]Ellis, Shine, Griffiths, Carlson, Pennar, and Ehrlich, "Who Pays for Peace," 65-70.

[25]M. Cetron and O. Davies, *The Great Job Shake-out* (New York: Simon and Schuster, 1988).

[26]For example, Oak Ridge and Los Alamos labs set up a high temperature superconductivity center that involved over 40 companies and universities. Also, the DOE signed a pact with the National Center for Manufacturing Sciences, a nonprofit consortium of more than 100 companies in Ann Arbor, MI. With industry-funded projects (at 50% level or higher), the DOE withholds publication of technical data for up to two years to provide firms with a head start in the markets.

[27]J. Braham, "A Pink Slip Among the Blueprints," *Machine Design* 11 (1991): 35-39.

[28]Kaplan, 70.

[29]Bell.

[30]Bell.

[31]J. Pfeffer, "A Partial Test of the Social Information Processing Model of Job Attitudes," *Human Relations* 33, No. 7 (1980): 457-476; R. Zussman, *The Mechanic of the Middle Class: Work and Politics Among American Engineers* (Berkeley: University of California Press, 1985).

[32]S. Cramer and M. Keitel, "Family Effects of Dislocation, Unemployment, and Discouragement," *Family Therapy Collections* 10 (1984): 81-93.

[33]T. Perry, "In Search of Peaceful Pastures," *IEEE Spectrum* (November 1989): 49-52.

[34]M. Berlye, *Your Career in the World of Work* (Indianapolis, IN: Bobbs-Merrill Co., 1975).

[35]R. Perrucci and J. Gerstl, *Profession Without Community: Engineers in American Society* (New York: Random House, 1969); Zussman.

[36]Perrucci and Gerstl.

[37]Zussman.

[38]L. Kahle, B. Poulos, and A. Sukhdial, "Changes in Social Values in the United States During the Past Decade," *Journal of Advertising Research* (February/March 1988): 35-41.

[39]Perrucci and Gerstl.

[40]R. Addison, "Grounded Interpretive Research: An Investigation of Physician Socialization," in M. Packer and R. Addison (Eds.), *Entering the*

Circle: Hermeneutic Investigation in Psychology (Albany: New York University Press, 1989).

[41]Ibid.; T. Mizrahi, *Getting Rid of Patients: Contradictions in the Socialization of Physicians* (New Brunswick, NJ: Rutgers University Press, 1986).

[42]R. Parsons, "Enlightened Management Must Lay a Strong Foundation in Training New Engineers," *Industrial Engineer* (April 1991): 34-38.

[43]L. Gomez-Mejia, D. Balkin, and G. Milkovich, "Rethinking Rewards for Technical Employees," *Organizational Dynamics* 18, No. 4 (1990): 62-70.

[44]H. Erlanger, "Social Reform Organizations and the Subsequent Careers of Participants: A Follow-Up Study of Early Participants in the OEO Legal Services Program," *American Sociological Review* 42 (1977): 233-248; H. Erlanger and D. Klegon, "Socialization Effects of Professional School: The Law School Experience and Student Orientations to Public Interest Concerns," *Law and Society Review* 13 (1978): 11-35.

[45]M. Rokeach, *Beliefs, Attitudes, and Values* (San Francisco: Jossey-Bass, 1968).

[46]Ibid.

[47]Braham, 35-39.

[48]L. Jones, "The Effect of Unemployment on Children and Adolescents," *Children and Youth Services Review* 10, No. 3 (1988): 199-215; R. Liem and J. Liem, "Psychological Effects of Unemployment on Workers and Their Families," *Journal of Social Issues* 44, No. 4 (1988): 87-105.

[49]R. Galdston, "Observations on Children Who Have Been Physically Abused and Their Parents," *American Journal of Psychiatry* 122 (1965): 440-443; D. Gill, *Violence Against Children: Physical Abuse in the United States* (Cambridge, MA: Harvard University Press, 1970); T. O'Brien, "Violence in Divorce-Prone Families," *Journal of Marriage and Family* 33 (1971): 292-298.

[50]A. Kinicki, "Personal Consequences of Plant Closings: A Model and Preliminary Test," *Human Relations* 38, No. 3 (1985): 197-212.

[51]Cramer and Keitel, 81-93.

[52]Braham, 35-39.

[53]Ibid.

[54]A. Zaleznik, G. Dalton, and L. Barnes, *Orientation and Conflict in Career* (Boston: Harvard University Press, 1970).

[55]Pfeffer, 457-476.

[56]S. Sedge, "A Comparison of Engineers Pursuing Alternate Career Paths," *Journal of Vocational Behavior* 27 (1985): 56-70.

[57]G. Latta, "Union Organization Among Engineers: A Current Assessment," *Industrial and Labor Relations Review* 35, No.1 (1981): 29-42; P. Meiksins, "Professionalism and Conflict: The Case of the American Association of Engineers," *Journal of Social History* 19, No.3 (1986): 403-

421; Perrucci and Gerstl; T. Trick, "AES—Another Attempt to Herd Cats?" *Engineering Education* (January/February 1991).

[58]Bell.

[59]D. Christiansen, "Fighting the New Wars," *IEEE Spectrum* 26, No.11 (1989): 25.

[60]S. Benson, "Managing the Trade-Defense Relationship," *The Bureaucrat* 19 (1990): 39-43.

[61]M. Badawy, "How to Succeed as a Manager: Pt. 3. Why the Switch from Engineer to Manager is Difficult," *Machine Design* 53 (1981): 91-95.

[62]Bell.

[63]E. Krau, "Commitment to Work in Immigrants: Its Functions and Peculiarities," *Journal of Vocational Behavior* 24, No. 3 (1977): 329-339.

[64]S. Kandebo, "Engineer Shortfall Still Seen, Despite Industry Doldrums," *Aviation Week and Space Technology* 134, No. 9 (1991): 62-63.

[65]K. Chen, "Reversing Sagging Precollege Skills in Mathematics and Science," *IEEE Spectrum* (December 1990): 44-48.

[66]S. Ramo, "National Security and Our Technology Edge," *Harvard Business Review* (November/December 1989): 115-120.

[67]Chen, 44-48; R. Ellis, "Engineering and Engineering Technology Degrees, 1990," *Engineering Education* (January/February 1991): 34-44.

[68]Ramo, 115-120.

[69]Chen, 44-48.

[70]J. Suran, "A View from the Other Side," *IEEE Spectrum* (September 1990): 52-54.

[71]Ibid.

[72]Office of Technology Assessment, *Educating Scientists and Engineers: Grade School to Grad School* (Washington, D.C.: U.S. Congress, 1988).

[73]Ellis, 34-44.

[74]J. Owen, "Images of the Manufacturing Engineer," *Manufacturing Engineering* (November 1989): 56-62.

[75]Kandebo, 62-63.

[76]D. Christiansen, "Open Questions on Defense," *IEEE Spectrum* 27 No. 11 (1990): 29.

[77]W. Rogerson, "Quality vs Quantity in Military Procurement," *The American Economic Review* 80 (1990): 83-92.

[78]Kaplan, 70.

[79]B. Santo, "Industry Seeking Strength in Mergers," *IEEE Spectrum* 26, No. 11 (1989): 64-66.

[80]R. Nimroody, "Star Wars' Brain Drain," *Financial World* 155, No. 11 (1986): 72.

[81]Santo, 64-66.

[82]P. Noaker, "How the Best Make It," *Manufacturing Engineering* (November 1989): 52-54.

[83]E. Dolnick, "DARPA-The Pentagon's Skunk Works," *Across the Board* 25, No. 4 (1988): 28-35.

[84]D. Hughes, "Defense Department Must Exploit Commercial Technology," *Aviation Week and Space Technology* 133, No. 26 (1990): 23-25.

[85]Ibid.

[86]J. Adam, "A Die-Hard Engineer in the White House," *IEEE Spectrum* 24 (March 1990).

[87]Rokeach, *Understanding Human Values*, 15.

[88]Ibid., 20.

[89]A. Maslow, *Motivation and Personality* (New York: Harper and Row, 1954).

[90]E. Mesthene, "How Technology Will Shape the Future," *Science* CLXI (1968): 135-143.

[91]Kahle, Poulos, and Sukhdial, 5-41.

[92]K. Alexander, "Scientists, Engineers and the Organization of Work," *American Journal of Economics and Sociology* 40, No. 1 (1981): 51-66; Perrucci and Gerstl; Zussman.

[93]A. Skolnick, *The Intimate Environment* (Boston: Little Brown, 1983).

[94]Meiksins, 403-421; S. Rynes, P. Tolbert, and P. Strausser, "Aspirations to Manage: A Comparison of Engineering Students and Working Engineers," *Journal of Vocational Behavior* 32 (1988): 239-253; Zussman.

[95]Kohn 1977; Skolnick 1983.

[96]Rynes, Tolbert, and Strausser, 239-253.

[97]Latta, 29-42; Meiksins, 403-421; Perrucci and Gerstl; Trick; Zussman.

[98]W. Melton, "Equality and Freedom: Exploring Social Values and Social Issues Among Engineering Students," *Michigan Academician* 14, No. 3 (1982): 273-283.

[99]Feather.

[100]Melton, 273-283.

[101]Feather.

[102]Perrucci and Gerstl.

[103]Feather.

[104]Melton, 273-283.

[105]G. Haworth, R. Povey, and S. Clift, "The Attitudes Towards Women Scale (AWS-B): A Comparison of Women in Engineering and Traditional Occupations with Male Engineers," *British Journal of Social Psychology* 25 (1986): 329-334.

[106]Ibid.; R. Kanter, "Some Effects of Proportions on Group Life: Skewed Sex Ratios and Responses to Token Women," *American Journal of Sociology* 82, No. 5 (1976): 965-990.

[107]L. Olson, "Sex Linked Values: Their Impact on Women in Engineering," *Social Science Journal* 14, No. 2 (1977): 89-102.

[108]Ibid.

[109]Ellis, 34-44.

[110]E. Layton, "Frederick Haynes Newell and the Revolt of the Engineers," *Midcontinent American Studies Journal* 3 (1962): 21-22; E. Layton, *The Revolt of the Engineers: Social Responsibility and The American Engineering Profession* (Cleveland: Case Western Reserve University, 1971); Meiksins, 403-421.

[111]A. Karr, "Labor Letter: Demand Climbs for Women and Minorities," *The Wall Street Journal* (18 June 1991).

[112]R. Kopelman, "Psychological Stages of Careers in Engineering: An Expectancy Theory," *Journal of Vocational Behavior* 10, No. 3 (1977): 270-286.

[113]Pfeffer, 457-476; Zussman.

[114]Zussman.

[115]Rokeach, *Understanding Human Values*, 15.

[116]G. Allport, *Pattern and Growth in Personality* (New York: Holt, Rhinehart, and Winston, 1961).

[117]M. Larson, *The Rise of Professionalism: A Sociological Analysis* (Berkeley, CA: University of Berkeley Press, 1977); Meiksins, 403-421.

[118]Pfeffer, 457-476.

[119]Kopelman, 270-286; Rynes, Tolbert, and Strausser, 239-253.

[120]L. Bailyn and J. Lynch, "Engineering as a Life-Long Career: Its Meaning, Its Satisfactions, Its Difficulties," *Journal of Occupational Behavior* 4 (1983): 263-283; Rynes, Tolbert, and Strausser, 239-253; Zussman.

[121]Sedge.

[122]L. Bailyn, "Ready, Set and No Place to Go," *Professional Engineer* 50 (1980): 43-46; Bailyn and Lynch, 263-283; F. Guterl, "Spectrum-Harris Poll: The Career," *IEEE Spectrum* 21 (1984): 59-63; Rynes, Tolbert, and Strausser, 239-253.

[123]H. Greenwald, "Scientists and the Need to Manage," *Industrial Relations* 17 (1978): 156-167; Rynes, Tolbert, and Strausser, 239-253.

[124]Alexander, 51-66.

[125]Sedge.

[126]J. Gutman, "A Means-End Chain Model Based on Consumer Categorization Processes," *Journal of Marketing* 46 (1977): 42-60.

[127]Rynes, Tolbert, and Strausser, 239-253.

[128] H. Gough and A. Heilbrun, *The Adjective Checklist Manual* (Palo Alto, CA: Consulting Psychologists Press, 1980); Sedge.

[129]Sedge.

[130]Schein.

[131]Rynes, Tolbert, and Strausser, 239-253.

[132]E. Raudsepp, "The Manager: to Be or Not to Be," *Hydrocarbon Processing* (July 1972): 116-121.

[133]Alexander, 51-66.; F. Goldner and R. Ritti, "Professionalism as Career Immobility," *American Journal of Sociology* 72, No. 5 (1967): 490.

[134]Alexander, 51-66; R. Kanter, *Men and Women of the Corporation* (New York: Basic Books, 1977).

[135]Sedge.

[136]Alexander, 51-66.

[137]Z. Shapira and T. Griffith, "Comparing the Work Values of Engineers with Managers, Production, and Clerical Workers: A Multivariate Analysis," *Journal of Organizational Behavior* 11 (1990): 281-282.

[138]Badawy, 91-95; J. Bayton and R. Chapman, "Professional Development: Making Managers of Scientists and Engineers," *The Bureaucrat* 1 (1977):408-425.; Zaleznik, Dalton, and Barnes.

[139]Kopelman, 270-286.

[140]Pfeffer, 457-476.

[141]K. Fitzgerald, "Whistle-Blowing: Not Always a Losing Game," *IEEE Spectrum* (December 1990): 49-52; W. Lowrance, *Modern Science and Human Values* (New York: Oxford University Press, 1986).

[142]J. Mingle and C. Reagan, "Legal and Moral Responsibilities of the Engineer," *Chemical Engineering Progress* 76, No. 12 (1980): 15.

[143]T. Kuhn, *Structure of Scientific Revolutions*, second edition (Chicago: University of Chicago Press, 1970).

[144]Ibid., 164.

[145]M. Raskin and H. Bernstein, "Toward a Reconstructive Political Science," in M. Raskin and H. Bernstein (Eds.), *New Ways of Knowing* (Totowa, NJ: Rowman and Littlefield, 1987).

[146]Zussman, 189.

[147]J. Christenson, J. Houghland, and B. Gage, "Value Orientations of Organized Religious Groups," *Sociology and Social Research* 68 (1984): 194-207.

[148]Rokeach, *Understanding Human Values.*; G. Sawa, *Reference Group Influence and Political Persuasion* (unpublished doctoral dissertation: Washington State University, 1985).

[149]M. Rokeach and J. Grube, *A Values Approach to Reducing and Preventing Adolescent Smoking* (unpublished manuscript, 1985).

[150]T. Greenstein, "Behavior Change Through Value Self-Confrontation: A Field Experiment," *Journal of Personality and Social Psychology* 34 (1976): 254-262.

[151]S. Sawa and G. Sawa, "The Value Confrontation Approach to Enduring Behavior Modification," *The Journal of Social Psychology* 128, No. 2 (1987): 207-215.

[152]S. Simon, L. Howe, and H. Kerschenbaum, *Values Clarification: A Handbook of Practical Strategies for Teachers and Students* (New York: Hart, 1972).

[153]Simon, *Meeting Yourself Halfway.*

[154]M. Heidegger, "Only a God Can Save Us," in *Heidegger: The Man and His Thought* (Trans W. Richardson) (Chicago: Precedent, 1981).

[155]Heidegger, 41.

[156] J. Sartre, *Being and Nothingness* (New York: Pocket Books, 1956).

[157]R. Bellah, R. Madsen, A. Sullivan, and S. Tipton, *Habits of the Heart* (New York: Harper and Row, 1985); A. Bloom, *The Closing of the American Mind* (New York: Simon and Schuster, 1987); F. Lappe, *Rediscovering America's Values* (New York: Ballantine Books, 1989); M. Wallach and L. Wallach, *Rethinking Goodness* (Albany: State University of New York Press, 1990).

[158]Wallach and Wallach.

[159]Bloom.

[160]Wallach and Wallach.

[161]P. Roubiczek, *Ethical Values in the Age of Science* (New York: Cambridge University Press, 1969).

[162]L.E. Grinter, "Report on Evaluation of Engineering Education, 1952-1955," *Journal of Engineering Education* 46, No. 1 (1955).

[163]D. Vines, "International Technical Education: A Comparative Study of Five Countries," *IEEE Technology and Society Magazine* (September/October 1990): 29-32.

[164]Zussman, 60.

[165]Ibid., 62.

[166]T. Reynolds and J. Gutman, "Laddering Theory, Method Analysis, and Interpretation," *Journal of Advertising Research* (February/March 1988): 11-23.

[167]R. Reich, "Who is Them," *Harvard Business Review* 69, No.2 (1991): 77-88.

[168]C. Morris, *The Coming Global Boom* (New York: Bantam, 1990).

[169]J. Dixon, "New Goals for Engineering Education," *Mechanical Engineering* (March 1991): 56-62.

[170]H. Knepler, "The New Engineers," *Change* 9 (1977): 30-35.

[171]Dixon, 56-62.

[172]J. Herkert, "Science, Technology, and Society Education for Engineers," *IEEE Technology and Society Magazine* (September/October 1990): 22-26.

5
Engineers' Perspectives on Defense Work and Economic Conversion

Eugene J. Vitale

Abstract

The funding, design, and manufacture of weapons systems by defense industries are based on a different set of criteria than for commercial products. Since the federal government provides much of the financing for weapons development, as well as purchases most of the product, the close linkage between the federal government and defense companies creates a production and marketing system unlike any in the commercial sector.

Successful economic conversion may rely on transferring technology from defense industries to civilian manufacturing; therefore, understanding the differences between the defense sector and the private sector is important if resources are to be successfully transferred. In this chapter the major differences between the two sectors are identified. Additionally, the results of a limited survey are discussed to provide insight into the different attitudes and values held by both defense and civilian engineers.

Introduction

One of the areas in which little research has been done is the values of engineers. This chapter reports the findings of a survey that focused on the value structure of engineers and their career paths. Specifically, it asked how engineers felt about themselves, their jobs, their employment, and economic conversion. This chapter reports the results of the survey and outlines differences between defense and nondefense contracting.

Roles of Government and Contractor

A major defense program with fully exercised growth potential can operate twenty to thirty years. For example, the Army's ground-to-ground missile, which was tube-launched, optically sighted, and wire-guided, went into production in the early 1960s and was finally used in the recent Gulf war. Programs such as these are initiated when the military identifies some need for a product. Thus, after the Soviets developed large army tanks, the U.S. military perceived the need for an antitank weapon. And because tanks are a threat to the Army, the Army was responsible for developing a statement of work and a request for proposals that it sent to viable contractors. These contractors in turn responded with product concepts and manufacturing plans. After the Army chose a contractor, the work proceeded from the developmental stages into production, and the product was ultimately delivered into the defense arsenal. The Soviets then countered by making tank armor thicker, and later, reactive. The U.S. Army responded each time by requesting proposals for countering product upgrades. Hence products mature, grow in sophistication, or expand in response to changes in the original threat. Table 1 outlines the phases of a major defense product and identifies the roles of the government and industry.

Program Funding

In government contracting, the contractor designs and develops a product for a known customer. This is very different from the commercial world, in which the manufacturer develops a product to meet a perceived need and then finds a customer.

Figure 1 shows how the government defrays the investment in product development but also limits the potential profit for defense contracting. In contrast a commercial venture requires full front-end investment but maintains a much higher potential for return on that investment until the market becomes saturated. One economic benefit that is often overlooked is how defense contractors provide work for major subcontractors and suppliers. Between 60% and 70% of a major defense system is purchased. Such procurements create a high degree of synergism in the commercial world, especially in the areas of manufacturing equipment and factory supplies.

Table 1

Program Phases Showing Government and Contractors Roles

MILITARY	CONTRACTOR
<u>Identifies a need</u>	<u>Writes a proposal</u>
Mission profile	program plan
reliability	product concept
maintainability	manufacturing plan
sustainability	logistics plan
testability	quality plan
cost	
program concept of	
manuscript and	
management	
select contractors	
request for proposals	
<u>Product demonstration and</u> <u>validation</u>	<u>Product demonstration and</u> <u>validation</u>
critique of demonstration or	demonstrates
validate	concept/product
chooses contractor	validates design
approves funding	proof of manufacturing
provides funding,	hardware
management, and test	deliverables
support	
selects subcontractors	
<u>Initiates full-scale development</u>	<u>Initiates full-scale development</u>
product analysis	proof of design hardware
provides funding, management	proof of manufacturing
and test support	hardware
selects subcontractors	deliverables
<u>Production</u>	<u>Production</u>
provides shipping instructions	provides hardware
specifies storage depot	produces manuals,
monitoring funding and	maintenance, and instruction
delivery	provides training
	manages cost, schedule, and
	quantity performance

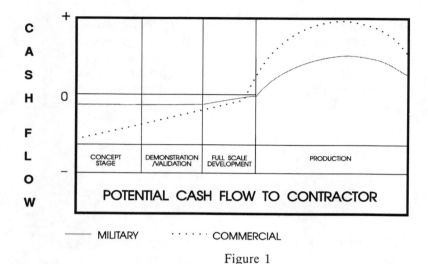

Figure 1

Opportunities for Conversion

Most defense contractors have commercial product lines and experience with non-defense business. These companies routinely review commercial ventures. Opportunities for conversion exist in government, commercial, and consumer sectors.

Other Government Sectors

Given that defense contractors are steeped in government contracting procedures, nondefense government business offers the best conversion opportunities. As discussed earlier, because government contractors build to a specification for a known customer, they do not have commercial marketing or distribution capability. Business opportunities within the Federal Aviation Administration, Post Office, Internal Revenue Service, Treasury Department, and other government agencies provide niches for defense technologies and support services. For example, missile technology uses pattern recognition to identify target signatures, something that can also be used in optical character recognition in postal equipment.

Commercial Sector

The most likely ventures for defense contractors in this sector will require teaming with commercial companies, such as Tandem or Fluor, that have established marketing and distribution networks. Commercial ventures may also emerge after evaluating smaller undercapitalized companies that have unique technologies in both product and production processes.

Consumer Sector

Expansion into new markets will most likely be from using existing or new technologies in commercial products. But to enter the consumer sector, sales skills, marketing expertise, and distribution networks are essential.

Commercial Diversification

One defense company with an aggressive approach to diversification toward commercial production is Hughes Aircraft. For Hughes, the difference in market development between defense and commercial markets involve (1) the need for internal definition of product requirements (unlike defense production where military specifications define products), (2) early entry in commercial markets, (3) systems design that produce payoffs of customers, (4) knowledge of commercial markets and customers, and (5) selection of markets where they have a unique sustainable competitive advantage (e.g., markets where Hughes owns patents in important technologies).

Survey Results

We sent our survey of engineers' reactions to conversion to 100 engineers in the west and midwest who are employed by a variety of manufacturing companies. The research design included a range of commercial ventures from "Fortune 500" to some "start-up" companies. The Fortune 500 group included appliance and home environmental control products as well as major defense contractors. An effort was made to reach male and female engineers representing a range of technical degrees and employment experience.

Eighty-nine of those who received questionnaires responded. Our sample included more women than is representative of the population of engineers. Further respondents differ from the total engineering workforce in that they were more likely to have masters degrees, to be managers or supervisors, and to work in the defense sector. The difference in defense employment is noteworthy: while 20% of all engineers nationally are estimated to be employed in the defense sector, 59% of our respondents report this type of work.[1] Because the sample was derived from professional contacts, those who responded may or may not be representative of engineers in manufacturing as a whole. The results are of interest in providing initial information about some attitudes and values of these engineers and for motivating more systematic research. Survey results can be found at the end of this chapter.

Career Issues

One interesting finding is that female and male engineers have similar ways of ranking the factors affecting their career choice. Job satisfaction and salary were ranked numbers one and two by all respondents and by women, who comprise 14% of the respondents. All engineers ranked defense, nondefense, societal problems, moral issues, and patriotism as the least important factors. Gender differences appeared for job recognition, job autonomy, and nondefense work which women rated higher, and societal problems, which men rated higher than the total population. See Tables 2 and 3 for a summary of career choice influences. Eighty-eight percent stated they were satisfied with their work when asked to rate the importance of job satisfaction, being a company decision maker, and being empowered (1 being high and 5 low). Of these factors, job satisfaction was rated the highest with a mean of 1.47. Seventy-six percent placed job satisfaction at the top.

Perceptions of Defense and Civilian Development

If conversion occurs and many engineers move from employment in the defense sector to the civilian sector, their perceptions of the characteristics noted earlier are key issues. For example, statistics show the defense sector is more lucrative than civilian work, and the ranking discussed earlier indicates that salaries are a prime motivator for engineers' career choice. If civilian employment is perceived to entail a significant reduction in pay, then engineers may resist conversion.

Respondents were asked to compare eight dimensions of defense and civilian work. As shown in Table 4, over half of these engineers believe that the two types of work are similar with respect to job recognition, company loyalty, job autonomy, and personal growth. The majority of respondents see the sectors as dissimilar in terms of technology and salary. For those who perceive a difference between the defense and civilian sectors, civilian work is seen as better with respect to job recognition, company loyalty, personal growth, job autonomy, job satisfaction, and job security. Defense work is seen more favorably with respect to salary and technology. Because salary is known to be an essential motivator, it may be an important determinant of engineers' reactions of the possibility of economic conversion. Similarly, engineers like technology.[2] This interest in technology may be a barrier to civilian work insofar as defense work is perceived to have better technology although this interest may be offset by the interest of experienced engineers moving into management.[3] These findings also suggest that engineers could be encouraged to support economic conversion by highlighting the advantages of civilian jobs in the civilian sector: job security, personal growth, recognition, autonomy, and job satisfaction.

Table 2
Career Choice: Female Engineers

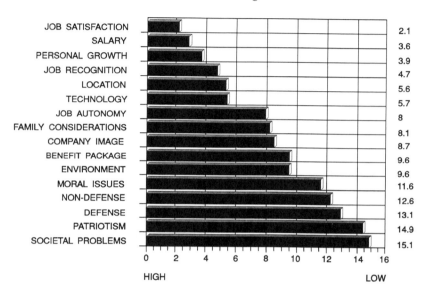

Table 3
Career Choice: All Participants

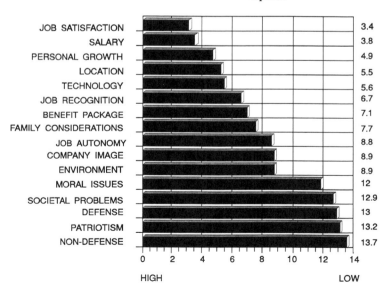

Table 4
Attitudes About Defense and Nondefense Employment

	BETTER IN		ABOUT THE SAME
	Defense	Nondefense	
Job recognition	12%	27%	61%
Company loyalty	18%	25%	57%
Personal growth	16%	31%	53%
Job autonomy	12%	35%	53%
Job satisfaction	16%	35%	49%
Job security	22%	35%	43%
Salary	41%	16%	43%
Technology	60%	12%	28%

Attitudes of Defense Engineers Toward Economic Conversion

Those engineers who were currently employed in a defense industry were asked what would happen if defense spending were to shrink. Only one-quarter of the respondents believed that they would continue to do defense work. Forty-three percent would stay with the same company and 77% think that their company could change to civilian work. Although nearly all of the engineers had thought about a shrinking defense budget (94%), uncertainty is evident with respect to their employment and the direction of their companies. The most common response as to whether they would stay with the same company or continue to do defense work was "don't know." Twelve percent did not know if their company could change to civilian production.

All respondents were asked to rate new directions for their companies. In their responses, engineers' business orientation is evident.[4] Engineers rated growth opportunities, success potential, and profitability as the most important factors. Issues with larger social implications—from rebuilding the infrastructure to environmental liability—received low ratings (see Table 5). The defense versus civilian work received the seventh-lowest rating, indicating that engineers feel little desire to shift out of defense work. That two-thirds of the respondents believe that professional engineering societies help develop values for engineering indicates the significant role that these societies can play in shaping engineers' role in economic conversion.

Can Engineers Help or Direct Conversion?

Engineers overwhelmingly believe they can help their companies make the transition and that they will be prime movers. They responded affirmatively 91% and 89% of the time, respectively, to "Could you help make the change?" and "Will engineers be instrumental in such a change?." (See question 6 of the survey.) This is particularly interesting in that based on their experiences to date, only 59% feel they are company decision makers and 69% feel empowered. These data suggest that given the opportunity, engineers would be instrumental in changing their companies' direction.

What Do Engineers Think About Defense Spending and the "Peace Dividend"?

Defense spending gets good marks from engineers. Seventy-eight percent felt that it had a positive impact on the U.S. economy. On U.S. technological leadership 88% rated it as a positive factor. But only 55% said that it had a positive impact on U.S. worldwide competitiveness. When it comes to reallocating the "peace dividend" engineers gave highest priority to U.S. technology, education, and recapturing U.S. manufacturing leadership (Table 6). They gave the lowest priority to social, environmental, and global issues. These priorities reflect engineers' practical orientation, their pro-business stance, and their conservative political orientation.[5]

Conclusion

Defense and civilian engineering differ significantly from an organizational viewpoint. Defense engineers design a product for a known customer in a situation that reduces the investment in product development and long term profit potential. Civilian engineers develop a product to meet a perceived need and then use marketing and sales personnel to find customers and distribute the product. The initial costs of product development are born by the company, but the potential profits are unrestricted. These differences will be salient for companies that convert from defense to civilian engineering and suggest that other government contracts or teaming with commercial companies will facilitate the transition of conversion.

    The survey of engineers reported here underscores the implication for engineers that the shift from defense to civilian engineering will bring. Salary is a driving force in job choice for engineers; insofar as civilian engineering pays less, engineers who convert are likely to be less satisfied with their jobs. On the other hand, civilian engineering is perceived to be better than defense work in terms of recognition, autonomy, personal growth, job satisfaction, and job security. These advantages may offset changes in salary in engineers' overall evaluation of their employment situation. Engineers are optimistic about the ability of their companies to convert to nondefense work, and nearly all believe that engineers generally and themselves specifically can be instrumental in making this change. That these engineers' orientation for new directions for their companies is focused upon business issues indicates that social, environmental, and global uses of the peace dividends are unlikely to occur insofar as conversion planning is directed by engineers and managers of existing firms.

Table 5
Defense Budget Reallocation

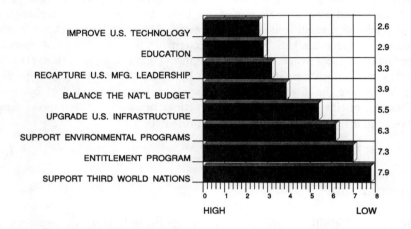

Table 6
Identifying New Directions

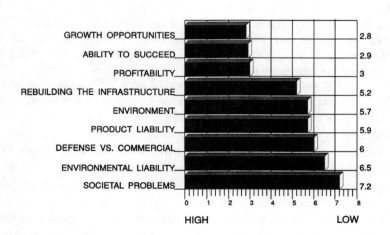

ENGINEERS AND ECONOMIC CONVERSION
SURVEY RESULTS JUNE 3, 1991

The University of Arizona is participating in a research program to gather data on engineers and economic conversion (transition from defense orientation to nondefense orientation). They wish to better understand the role of engineers in industry and their values and attitudes. To that end, please take a few minutes and complete the following questionnaire. Be assured that your responses will remain anonymous.

1)   PRIORITIZE THE FOLLOWING FACTORS FROM 1 TO 16 AS TO HOW THEY INFLUENCED YOUR CAREER CHOICE. 1 = HIGHEST AND 16 = LOWEST

Salary	2
Defense orientation	14
Nondefense orientation	16
Societal problems	13
Technology	5
Family considerations	8
Location	4
Patriotism	15
Moral issues	12
Benefit package	7
Environment	11
Job autonomy	9
Company image	10
Job satisfaction	1
Personal growth	3
Job recognition	6

2)   DO YOU CONSIDER YOURSELF TO BE AN (CHECK ONE)

Engineer	51%
Technical supervisor	8%
Manager	41%

2A)   WHAT IS / WAS YOUR PRIMARY RESPONSIBILITY?

Design	28%
Manufacturing engineering	22%
Production support	20%
Planning	____
Administration	6%
Product development	11%
R & D	8%
Marketing	5%
Other	____

3)   WHAT HAS BEEN YOUR WORK EXPERIENCE?

Dominantly defense	59%
Dominantly nondefense	31%
About equal	10%

4)   BASED UPON YOUR EXPERIENCE OR EXPECTATIONS, CHECK THE BOX THAT BEST APPLIES FOR EACH OF THE FOLLOWING CATEGORIES.

	BETTER IN		ABOUT THE SAME
	Defense	Nondefense	
Job satisfaction	16%	35%	49%
Job security	22%	35%	43%
Company loyalty	18%	25%	57%
Salary	41%	16%	43%
Personal growth	16%	31%	53%
Technology	60%	12%	28%
Job recognition	12%	27%	61%
Job autonomy	12%	27%	61%

5)   BASED ON YOUR EXPERIENCE CHECK YES OR NO AND RATE

	YES	NO	IMPORTANCE TO YOU 1 = High 5 = Low
Satisfied with your work	88%	12%	Mean = 1.47
A company decision maker	59%	41%	Mean = 2.33
Empowered	69%	31%	Mean = 1.78

6)   IF YOU ARE CURRENTLY EMPLOYED IN A DEFENSE INDUSTRY, ANSWER THE FOLLOWING. IF DEFENSE SPENDING SHRINKS...

	Yes	No	Don't know
Will you stay with same company?	43%	11%	46%
Will you try to stay in defense work?	26%	37%	37%
Could your company change to commercial work?	77%	11%	12%
Could you help make the change?	91%	9%	—
Will engineers be instrumental in such a change?	89%	11%	—
Had you thought about it?	94%	6%	—

7)   DO ENGINEERS PARTICIPATE IN DECISIONS THAT CHANGE YOUR COMPANY'S DIRECTION?

Yes	No	Don't know
71%	27%	2%

8)   IF YOU WERE IDENTIFYING NEW DIRECTIONS FOR YOUR COMPANY, HOW WOULD YOU RANK THE FOLLOWING? INDICATE 1 THROUGH 9.

Growth opportunities	1
Societal problems	9
Environment	5
Product liability	6
Environmental liability	8
Rebuilding the infrastructure	4
Profitability	3
Ability to succeed	2

9)   WHAT IS THE IMPACT OF DEFENSE SPENDING...

	Positive	Negative	None
On the U.S. economy?	78%	22%	—
On U.S. worldwide competitiveness?	55%	37%	8%
On U.S. technological leadership?	88%	6%	6%

10)   ASSUMING THE DEFENSE BUDGET WILL CONTINUE TO SHRINK, HOW WOULD YOU REALLOCATE THE MONEY? INDICATE 1 THROUGH 8.

Balance the national budget	4
Support environmental issues	6
Entitlement programs	7
Upgrade U.S. infrastructures	5
Support third-world nations	8
Improve U.S. technology	1
Recapture U.S. manufacturing leadership	3
Education	2

11)   FROM A TECHNICAL PERSPECTIVE IS THERE A SIGNIFICANT DIFFERENCE BETWEEN NONDEFENSE AND DEFENSE CAREERS?

Yes	No
56%	44%

12)   DO YOU BELIEVE THAT PROFESSIONAL ENGINEERING SOCIETIES HELP
       DEVELOP VALUES FOR ENGINEERS?

Yes	No
65%	35%

13)   INDICATE THE HIGHEST LEVEL OF EDUCATION THAT YOU COMPLETED.

High school	2%
Associates	2%
Bachelors	53%
Masters	39%
Doctorate	4%

14)   INDICATE YOUR AGE AND GENDER.

Age	Male	Female
20-24	4%	—
25-34	24%	8%
35-44	28%	4%
45-54	14%	2%
55 plus	16%	—

Notes

[1]See Yudken and Markusen chapter in this volume.

[2]R. Perrucci and J. Gerstl, *Profession Without Community: Engineers in American Society* (New York: Random House, 1969); S. Rynes, P. Tolbert, and P. Strausser, "Aspirations to Manage: A Comparison of Engineering Students and Working Engineers," *Journal of Vocational Behavior* 32 (1988): 239-253.

[3]Rynes, Tolbert, and Strausser (1988) compared managerial career anchors and technical career anchors. They found the engineering students were more tied to the managerial anchor (policy making, salary, promotion) but that experienced engineers were more interested than students in consulting, starting businesses, and entering academics.

[4]M. Larson, *The Rise of Professionalism: A Sociological Analysis* (Berkeley, CA: University of Berkeley Press, 1977); P. Meiksins, "Professionalism and Conflict: The Case of the American Association of Engineers," *Journal of Social History* 19, No.3 (1986): 403-421.

[5]Ibid.

6

National Technology Priorities and Economic Conversion

Kent H. Hughes

Abstract

The steady erosion of U.S. technological prowess over the last ten years is leading to a lack of competitiveness for U.S. industry. U.S. firms are losing market share to other nations in technologies they once dominated, such as the production of memory chips and printed circuited boards. This situation, however, does not exist for all technologies; the U.S. is still a leader in numerous critical technologies, such as genetic engineering, systems engineering, and artificial intelligence. These technologies remain competitive for at least two reasons: (1) U.S. industry has invested heavily in their development and (2) government policies support these industries.

National technological priorities must change if the U.S. is going to gain a competitive edge in future technologies and regain market share in losing technologies. Government support for U.S. industry is needed for these to happen.

The possibility of cuts in defense spending may leave many engineers at government-supported national labs unemployed. The skills of these engineers could be used in the economic conversion process to transfer defense technology to the civilian sector. Redirecting the resources from the national labs may also increase the competitiveness of U.S. industry abroad by creating new products for sale in foreign markets.

Past attempts at technology transfer between the labs and private industry have not been successful for at least three reasons: (1) a legal environment in which laws restrict federally funded research in the labs and provide few incentives to create consumer-oriented technology; (2) organizational and interorganizational structural problems that make cooperative research between the labs and the private sector difficult; and (3) differing organizational cultures related to differences in design parameters and value systems. Overcoming the problems between the national labs and industry will be difficult. However, recent legislation eliminates some of the legal barriers, and the exchange of information between the labs and industry is increasing the number of cooperative development efforts. The leadership in industry and the national labs must make cooperation in technology development and transfer a priority. Without a refocusing of U.S. technological priorities, U.S. competitiveness may continue to decline.

Critical Period in U.S. Technology History

U.S. technological preeminence is at risk. The Council on Competitiveness, in its recent report *Gaining New Ground: Technology Priorities for America's Future*, concluded, "The strong across-the-board U.S. position of a decade ago has deteriorated significantly. U.S. industry has already lost several technologies that are critical to industrial performance, and it is weak or losing badly in others. Moreover, the trends are running against the United States—in most technologies, the U.S position continues to erode."[1]

Similar conclusions have been reached by several other groups that have assessed American technological capability, including the Computer Systems Policy Project, the National Academy of Engineering, the MIT Commission on Industrial Productivity, the Department of Commerce, the Defense Science Board, and more recently, the National Advisory Committee on Semiconductors, the White House Office of Science and Technology Policy, and the Congressional Office of Technology Assessment. These groups agree that certain technologies—variably described as "critical," "emerging," and "pre-competitive, generic enabling"—are of great importance to America's competitive future, and in several of them there is widespread recognition that the United States has become uncompetitive.

The Evidence on U.S. Technologies: How We Measure Up

The Council on Competitiveness report, *Gaining New Ground*, represents the first-ever U.S. private-sector consensus on U.S. priorities in technology. The critical technologies cut across nine major sectors of the economy, which together account for $1 trillion in sales. The report describes where the United States is ahead, where it is behind, and, more important, what U.S. government, labor, industry, and universities can do to improve America's performance.

Technologies in Which the U.S. Position is Strong

The Council report found that technologies where the United States is strong are generally areas where the private sector has aggressively invested in technology and where the U.S. policy environment has been supportive. The technologies in which the United States shows strength are shown in Table 1.

Table 1

Technologies in Which the United States is Strong [2]

MATERIALS AND ASSOCIATED PROCESSING TECHNOLOGIES
  Bioactive/biocompatible materials
  Bioprocessing
  Drug discovery techniques
  Emissions reduction
  Genetic engineering
  Recycling/waste processing
ENGINEERING AND PRODUCTION TECHNOLOGIES
  Computer-aided engineering
  Systems engineering
ELECTRONIC COMPONENTS
  Magnetic information storage
  Microprocessors
INFORMATION TECHNOLOGIES
  Animation and full motion video
  Applications software
  Artificial intelligence
  Computer modeling and simulation
  Data representation
  Data retrieval and update
  Expert systems
  Graphics hardware and software
  Handwriting and speech recognition
  High-level software languages
  Natural language
  Neural networks
  Operating systems
  Optical character recognition
  Processor architecture
  Semantic modeling and interpretation
  Software engineering
  Transmitters and receivers
POWERTRAIN AND PROPULSION
  Airbreathing propulsion
  Low-emissions engines
  Rocket propulsion

The Council report notes several reasons for the strength in these technologies:

(1) The strong U.S. position in biotechnology is directly related to the combination of three elements: the intensity of the U.S. pharmaceutical industry's commitment to research; significant federal support of biological research, particularly by the National Institutes of Health (NIH); and the relatively unfettered flow of information between government-supported and industry researchers. U.S. government-supported research, which greatly exceeds that by any other country, contributed substantially to the creation of modern biotechnology.

(2) The favorable U.S. position in several environmental technologies is traceable to the strong demand for these technologies created by domestic regulations. Although this demand creates an advantage for U.S. companies exploiting these technologies, companies from other countries are also moving quickly to meet the demand.

(3) The strong U.S. position in a number of computer-related technologies reflects the historically powerful U.S. computer industry, which was supported in its early days by Department of Defense (DoD) research and procurement and sustained its lead due to leading-edge U.S. university and computer industry research. A climate that supports entrepreneurship has also aided the U.S. position in many of these technologies.

(4) The strong U.S. position in propulsion technologies is largely due to steady and significant federal procurement of products using these technologies, experience in systems expertise, high barriers to entry for newcomers to the field and significant investments in aerospace and defense research.[3]

The report cautions, however, that even though the United States remains strong in these technologies, there is no cause for complacency. "In many of these technologies, the U.S. lead is diminishing, and in virtually no cases is the U.S. lead significantly increasing."[4]

Technologies in Which the U.S. Position is Competitive

The Council drew up a second list for those technologies in which the United States is still "competitive" or roughly on par with the best of the world. The Council found no clear pattern to the technologies in which the United States is competitive. These technologies include those, such as logic chips and photoresists, in which the United States once dominated but others have caught up. They also include technologies, such as optical materials, in which different countries lead in different market niches. Other technologies, such as superconductors and submicron technology, are still in the research stage, and no clear leader has been established. The report finds that a number of these technologies are integral to a few large U.S. industries, such as chemicals, automobiles, and telecommunications, and the resources of these industries help keep the United States competitive in the underlying technologies. But, here too, the trends are not encouraging.[5]

Table 2

Technologies in Which the United States is Competitive[6]

MATERIALS AND ASSOCIATED PROCESSING TECHNOLOGIES

      Catalysts
      Chemical synthesis
      Magnetic matrix composites
      Net shape forming
      Optical materials
      Photoresists
      Polymers
      Polymer matrix composites
      Process controls
      Superconductors

ENGINEERING AND PRODUCTION TECHNOLOGIES

      Advanced welding
      Computer-integrated manufacturing
      Human factors engineering
      Joining and fastening technologies
      Measurement techniques
      Structural dynamics

ELECTRONIC COMPONENTS

      Logic chips
      Sensors
      Submicron technology

INFORMATION TECHNOLOGIES

      Broadband switching
      Digital infrastructure
      Digital signal processing
      Fiber-optic systems
      Hardware integration
      Multiplexing
      Spectrum technologies

POWERTRAIN AND PROPULSION

      Alternative-fuel engines
      Electrical storage technologies
      Electric motors and drives

Technologies in Which the U.S. Position is Weak

The Council report finds that many of the areas where the United States is weak reflect the effects of high capital costs; the lack of cooperative relations between equipment, materials, and component suppliers and their customers; and an underemphasis on manufacturing.[7] The report identified several features of the technologies in which the United States is weak:[8]

(1) Many of the materials and electronic component technologies are capital-intensive and characterized by rapid product cycles and pronounced business cycles. In several cases, the U.S. industry is made up of many autonomous, often smaller companies that have limited financial resources and have historically not cooperated with upstream or downstream companies to share risks in R&D. As a result, they have been more vulnerable to business cycles than their overseas competitors and have been less able to sustain investments in technologies.

(2) Weakness in many engineering and production technologies reflect a pervasive underemphasis on manufacturing in the United States. Many U.S. industries have lost market share because they were unable to produce as high a quality and as reliable a product as their competitors.

(3) U.S. manufacturers also have failed to lead in developing new manufacturing technologies and have failed to use these new technologies when they have been developed.

(4) The private sector has underinvested in R&D and slowed increases of R&D spending. While industry spending for R&D increased by about 27% in constant dollars in the first half of the 1980s, in the second half of the same decade it increased by only 5%.[9]

(5) The government role as a precursor in defense applications has been superseded by commercial sector first use. The old defense-to-commercial sector technology transfer model is today much less effective than in the past.

Technologies in Which the U.S. is Losing Badly or has Lost

Gaining New Ground found that the areas in which the United States is losing badly or has lost show similar characteristics to those in which the United States is weak: high capital costs, industry fragmentation, and an underemphasis on manufacturing. In addition, many of the areas where the United States has lost its standing are areas where concentrated foreign efforts have hurt the competitiveness of U.S. industry. For example, several technologies—such as microelectronics and their upstream and downstream support industries, as well as advanced materials—are ones that other countries have identified as critical and have supported in a focused way, using subsidies and government procurement, as well as trade and investment policies.   However, even in areas of overall weakness, the United States retains many strengths. For example, although U.S. industry is in general losing badly in memory chips, a few companies have substantial in-house capabilities, and U.S. merchant firms are successful in some niche products, such as erasable programmable read only memories (EPROMs) and static random access memories (SRAMs). There is also no doubt that the U.S. semiconductor industry is leading in the key area of microprocessors.

Table 3

Technologies in Which the United States is Weak[10]

MATERIALS AND ASSOCIATED PROCESSING TECHNOLOGIES

Advanced metals
Membranes
Precision coating

ENGINEERING AND PRODUCTION TECHNOLOGIES

Design for manufacturing
Design of manufacturing processes
Flexible manufacturing
High-speed machining
Integration of research, design, and manufacturing
Leading-edge scientific instruments
Precision bearings
Precision machining and forming
Total quality management

ELECTRONIC COMPONENTS

Actuators
Electro photography
Electrostatics
Laser devices
Photonics

POWERTRAIN AND PROPULSION

High fuel-economy/power density engines

Table 4

Technologies in Which the United States is Losing Badly or has Lost[11]

MATERIALS AND ASSOCIATED PROCESSING TECHNOLOGIES

Display materials
Electronic ceramics
Electronic packaging materials
Gallium arsenide
Silicon
Structural ceramics

ENGINEERING AND PRODUCTION TECHNOLOGIES

Integrated circuit fabrication and test equipment
Robotics and automated equipment

ELECTRONIC COMPONENTS

Electroluminescent displays
Liquid crystal displays
Memory chips
Multichip packaging systems
Optical information storage
Plasma and vacuum fluorescent displays
Printed circuit board technology

Why a Strong Position in Technology Matters

Over the past decade in particular, many studies have attempted to fix a quantitative value on the relationship between investment/development of new technology and productivity and economic growth.

In 1979, the Department of Labor found the following percentage contributions of various inputs to productivity: labor quality 10-18%, new technology 44-62%, and capital plant and equipment 18-42%.[12]

In a study of the economic performance of five industrial countries, Michael Boskin and Lawrence Lau found that technological progress is capital-augmenting and that the benefits of technological progress and economic performance thus complement each other. Together, they account for about 70% of the growth in output in the United States according to Boskin. [13] The point is not to establish exactly how much technological innovation affects productivity and economic growth. The fact is that it does, and to a large degree. The Council on Competitiveness' recent technology report was prepared because of the strong conviction of leading American private-sector executives (industry, labor, and academia) that "unless the nation acts immediately to promote its position in critical generic technologies, U.S. technological competitiveness will erode further, with disastrous consequences for American jobs, economic growth, and national security."[14]

U.S. policymakers have also gradually come to recognize through legislation the significance of technological innovation to national economic growth and security. As part of a law passed in 1989, Congress agreed "that technology advancement is a key component in the growth of the United States industrial economy and a strong industrial base is an essential element of the security of this country." The legislation further noted a "need to enhance U.S. competitiveness in both domestic and international markets....Innovation and the rapid application of commercially valuable technology are assuming a more significant role in near-term marketplace success."[15]

What Can Be Done to Improve the U.S. Technology Position?

Council Policy Recommendations

In *Gaining New Ground*, the Council on Competitiveness makes several specific recommendations on how to improve U.S. technological competitiveness. The five key general recommendations are listed here:

(1) In order to enhance U.S. competitiveness, the President should act immediately to make technological leadership a national priority.
(2) The federal and state governments should develop policies and implement programs to assure that America has a world-class technology infrastructure.
(3) U.S. industry should establish more effective technology networks to help it compete in the international marketplace.
(4) U.S. firms should set a goal to meet and surpass the best commercialization practices of their competition.
(5) While keeping their basic research programs strong, universities should develop closer ties to industry so that education and research programs contribute more effectively to the real technology needs of the manufacturing and service sectors.

Cold War to Conversion:The Chance to Change National Priorities

For more than forty years, a large share of U.S. public and private financial and technical resources have been focused on containment and deterrence. The sharp reduction in superpower hostilities in the past few years is good news both in the political sense and because of its potential benefit to U.S. economic competitiveness. Indeed, the defense debate in the United States has turned to the questions of how much to reduce defense spending and on what the so-called peace dividend should be spent.

Even before the reduction in superpower tensions, there were growing economic pressures on American defense budgets. Beyond the impact of the fiscal deficit, the emphasis on domestic needs and the impact of economic strength on everything from the defense industrial base to international standing was growing.

In recent years, the impact of defense technology on the commercial base has been less decisive. Increasingly, more defense research dollars have been spent on weapons testing and development, activities that offer less civilian potential than basic research or the development of generic technologies. Furthermore, in more and more cases, the leading edge is found in commercial rather than military technologies. Finally, the older model of inventing a military technology, identifying a commercial application, and bringing a product to market has simply become too slow to match the pace of international competitors that focus directly on developing commercial technology and marketable products.

The prospect of peace and the opportunity to shift resources to match new priorities offer enormous potential. But like much change, it can be wrenching for workers, engineers, and company executives who have to adapt existing skills and strengths to a new world. There are two models for "conversion." One calls for serious downsizing of firms, closing of military bases, and the movement of many highly skilled engineers to new jobs, even new careers, and/or new parts of the country. The second model calls for shifting the activities of the firm, rather than downsizing. This type of conversion has not been particularly successful in the private sector to date. However, in the case of some of the country's public sector activities, the "shifting activities model" deserves another look. Given the size and richness of the public resources involved, the country should not miss the opportunity to channel these resources toward meeting new challenges.

Economic Conversion and the Role of the National Labs

I will direct the remainder of this chapter to the question of technology transfer as it pertains to the national laboratory system and its relationship to industry. Currently over 800 national labs[16] that have such diverse research missions as agriculture, defense, energy, environment, and health exist. Many of these labs feature excellent research facilities and highly skilled technical staffs. Not to tap into these labs as a way to improve the nation's technological capability and potential—especially given the current level of foreign technological competition—would seem to be a serious misallocation of resources.

Yet, as suggested earlier, skepticism abounds about the ability of labs to transfer technology to the commercial sector. One of the Council on

Competitiveness' specific policy recommendations suggested that a strategy should be adopted to "make sure that the federal laboratories' contribution to national technology needs is commensurate with the national investment in them....Although the nation spends over $20 billion on the federal labs, their current culture and direction do not adequately support technology development that strengthens national economic performance."[17] The report also found that "much of U.S. industry is convinced from first-hand experience that the labs are not closely integrated with the private sector and therefore doubts that sustained investments in them will result in a significant payoff for generic industrial technology."[18]

The Council report notes that it may be necessary to "close and consolidate obsolete federal laboratories and scale back their funding in favor of university research." The Council goes on to make several positive suggestions that would help change negative perceptions about the labs, including greater private-sector involvement in the labs and a reallocation of resources to basic research in commercially relevant technologies. For example, the National Institutes of Health (NIH) is asked to strengthen research and enhance private-sector involvement in its programs, and the National Aeronautic and Space Administration (NASA) labs are asked to reallocate resources to emphasize basic and applied research and validation of technical disciplines that are important to commercial aircraft.[19]

The Lab-to-Industry Technology Transfer Process

Why the Skepticism?

Historically, three essential problems inhibited the lab-to-industry technology transfer:

The legal environment: Until the 1980s, regulation and laws restricted federally funded research conducted in national labs, especially in such areas as intellectual property, cooperative research, licensing, and researcher participation in commercial ventures.

Organizational and interorganizational structures: Only in recent years has serious thought been given to developing models for cooperative research and structures for research labs to work with industry on proprietary projects. Prior to the 1980s, an understanding of how to bridge the gap between research-performing institutions and industrial clients or partners was limited.

Differences in organizational cultures: Because they have two very different sets of design optimal parameters and operating constraints, researchers in labs and their technical counterparts in industry have traditionally functioned with very different perspectives and values. This divergence in approach includes different views of the research agenda;

different viewpoints on cost, volume, expectations, schedules, and the concept of "deliverables"; and different optimization goals for generic technologies.

Changes in the Law

In the first area—legal changes—much progress has been made over the past ten years. Congress has recognized through legislation the need for greater interaction between the national labs and industry.

> ...The Federal laboratories and other facilities have outstanding capabilities in a variety of advanced technologies and skilled scientists, engineers and technicians who could contribute substantially to the posture of U.S industry in international competition.... Improved opportunities for cooperative research and development agreements between contractor-managers of certain federal laboratories and the private sector in the United States, consistent with the program missions at those facilities, particularly the national security functions involving atomic energy defense activities, would contribute to our national well-being....More effective cooperation between those laboratories and the private sector in the United States is required to provide speed and certainty in the technology transfer process.[20]

Legislative action on economic conversion started with the Stevenson-Wydler Act (1980), which gave national laboratories the specific mission of transferring new technologies and required them to establish Offices of Research and Technology Applications (ORTAs) to effect the transfers. Successive acts of Congress, including the Federal Technology Transfer Act (FTTA, 1986) and the National Competitiveness Technology Transfer Act (NCTTA, 1989) have amended and extended the original provision. There have also been changes in internal policy at virtually every major research-performing institution in the United States on issues of intellectual property, participation of researchers in commercialization activities, and the internal regulations that pertain to commercially oriented research.[21]

Changes in Organizational Structures

Organizational and interorganizational structures that promote technology transfer have been put into place, particularly through the Federal Technology Transfer Act of 1986 and the National Competitiveness and

Technology Transfer Act (1989). These laws authorize government-owned, contractor-operated laboratories to participate in cooperative research and development agreements (CRADAs) with industry and call for enhanced collaboration between universities, the private sector, and government-owned, contractor-operated laboratories in order to foster the development of technologies in areas of significant economic potential.[22] To further encourage technology transfer and the commercialization of federal R&D results, the Technology Transfer Act also requires lab departments to create awards and royalty-sharing programs for federal scientists, engineers, and technicians. The technology transfer legislation furthermore provides for an exchange program for scientists and engineers between the private sector and federal laboratories.[23]

More recently, a provision relating to small business was incorporated into the 1991 Defense Authorization Act (the Small Business-National Defense Laboratory Partnership Act of 1990). This provision recognizes that "intermediaries" between the labs and small businesses help form and cement these partnerships. These intermediaries include state and local technological or economic development agencies, such as small business incubation centers, local colleges, and local economic development centers.[24] Senator Jeff Bingaman (D-NM), who sponsored the provision, sees it as playing a role in nudging small business, state and local government, and the labs along in productive cooperation.[25]

Model programs for the Small Business/Partnership Act might arrange for (1) meetings between small businesses and laboratory employees; (2) access to nonclassified work in process or completed at the laboratory; (3) use of laboratory facilities and equipment including software, hardware, processing technologies, and instruments; (4) license to patents and other intellectual property rights; and (5) adaptable engineering or manufacturing extension services. The laboratories could also enter into cooperative development agreements and engage in personnel exchanges with small businesses. Workshops, conferences, and similar meetings could be planned in order to disseminate information about the technology-related assistance available at the laboratory.[26]

In addition, the Department of Commerce was given a major role in coordinating the government's efforts to encourage commercial use of federal R&D by the private sector, via the Interagency Committee for Federal Laboratory Technology Transfer (set up in 1987 by former Commerce Secretary Malcolm Baldrige). Comprised of Assistant Secretary-level representatives from all federal agencies and departments in R&D, the committee addresses issues confronting departments and agencies in implementing the FTTA and contributes to the development of policies involving broader technology transfer issues. The committee also works closely with the newly established Working Group on Commercialization of Government Research (set up by the President through the Vice President's Council on Competitiveness).[27]

Deborah Wince-Smith (Department of Commerce) has stated, "There has been a sharp increase in the number of active cooperative research and development agreements between Federal agencies and departments and the private-sector—from 33 in FY 1987 to more than 450 in FY 1990. This demonstrates dramatic progress in getting industry to take advantage of the unique capabilities within our Federal laboratories."[28]

Changes in Organizational Cultures

Although the United States has made great progress over the past decade in eliminating some of the legal and structural constraints that had historically thwarted lab-to-industry technology transfer, general agreement persists that the differences between organizational cultures in research labs and in industry continue to impede the technology transfer process.

In recent testimony before Congress on the subject of technology transfer, one witness testified: "In the present culture, competent, highly motivated [lab] researchers are focused on a scientific or mission objective, and are not aware of commercial implication of their work....The researcher must become comfortable with the notion that there are potential non-mission-related applications of his/her work, and must be made aware of the needs that exist in industry that might be addressed by this work."[29]

Even a relatively optimistic administration official admitted that "while the legal framework is now in place for industry-government collaboration in R&D, there are still barriers to the success of such activities, many of which are 'cultural' in nature."[30] The same government official concluded that "attitudes both within the Federal laboratories and in industry must change if [the United States] is to derive maximum benefit from [its] Federal R&D investments."[31]

Some real differences in the way research institutions and industry view technology development remain primarily because they have two very distinct goals in mind. Industry is driven by product demands/requirements and constrained by considerations of cost, volume, and cycle time, whereas research labs are driven by the underlying knowledge pursued or by a specific mission requirement. Because the operating constraints (and consequently, the value systems) for each sector differ considerably, it is perhaps inevitable that their respective approaches to problem solving and technology development differ significantly. Each sector maintains that it is performing in an optimal way, given its boundaries and missions.

In the case of the federal laboratories and private industry, a half century of deliberate legal and structural blocks on interaction between the two sectors has further worked to harden the cultural differences between research labs and industry. Because of these entrenched differences, recent legislation that has sought to encourage more interaction between the two sectors is only slowly having its intended impact.

Effectiveness of Technology Transfer

GAO Study

In a recent study, which examined the national lab system through 1989, the General Accounting Office concluded that the federal labs have been doing a very uneven job of implementing the Federal Technology Transfer Act.[32] A senior GAO official noted in recent testimony before the House Subcommittee on Technology and Competitiveness, "We found that although the federal departments that we contacted had taken steps to implement the technology transfer legislation and the executive order, none had achieved full implementation."[33] He described the Environmental Protection Agency (EPA), the Department of Energy (DOE), and the Department of Transportation (DOT) labs as having the highest levels of implementation and the Department of Interior and Veterans Affairs (VA) as having the lowest. The labs were assessed based on (1) received implementation guidance from headquarters; (2) received authorization to enter into cooperative R&D agreements; (3) established Offices of Research and Technology Applications, if required; (4) staffed the ORTAs with at least one full-time position, if required; and (5) established awards, royalty-sharing, and personnel exchange programs.[34]

ITSR Study

The Institute for Technology and Strategic Research (ITSR) has also studied the technology transfer process, focusing on the Department of Energy defense labs.[35] The authors of the resulting report noted that more than a year after the NCTTA became law, the proposed CRADA program had not been implemented. They blamed such bureaucratic problems as too much time spent on arriving at an operating draft of the guidelines, the delayed approval of contract modifications by operating contractors, and the delay caused by labs not being permitted to go forward under pre-NCTTA frameworks for cooperative research. The ITSR research team determined that too many people had been brought into the process. The ITSR team recommended, among other things, that the Technology Transfer Project Group, the Technology Transfer Field Task Force, and the Technology Transfer Steering Committee of Defense Programs be terminated in favor of a more comprehensive DOE implementation plan.

The ITSR analysts cautioned that the new Advanced Manufacturing Initiative (AMI) should not be burdened with the same time-consuming, multi-layered controls that attended the birth of the CRADA program. The ITSR report notes that various federal agencies have established technology transfer programs, and the models that should be looked at for closer study are those that go beyond patent licensing and afford significant opportunities for cooperative research with industry, most notably the National Institute of

Standards and Technology (NIST) and NASA, as well as a contractor-operated lab, the Jet Propulsion Lab (JPL). The ITSR research team stressed the fundamental importance of two-way communication between national lab and industrial scientists.[36]

ITI Project

Because of the current lack of information and interaction between lab and industry, the Industrial Technology Institute (ITI) has developed a needs-driven approach to technology transfer that calls for using a facilitator or intermediary between lab and industry.

ITI produced a 250-page book organized by ten durable goods industries and SIC manufacturing functions that defines the technology needs in each industry sector surveyed. With the assistance of lab management, ITI staff identifies a group of 15-20 individuals who have both active research programs and an interest in a more ambitious program of technology transfer; these researchers are then given the ITI book. ITI staff meet with lab managers to raise the perceived importance of *needs* as the primary focus for identifying commercialization possibilities, and together they identify a list of 15 to 20 candidate technologies. ITI staff and technical experts then screen the nominated technologies, narrowing the list to two to four technologies, based on significant potential for commercialization.

For those technologies selected, the staff develops minibusiness plans for each. This involves ITI analysis of prior art and market research; the identification of potential commercial partners; and an understanding of any need for advanced development prior to transfer. Sometimes additional technology development is needed, because many of the inventions that are nominated are in fact not sufficiently "hardened" to pique the interest of a potential partner. This is particularly so in durable goods manufacturing. The final step is finding a commercial partner and negotiating a commercial agreement. ITI has data bases of potential commercializers and has developed a disciplined approach for the search, qualification, and negotiation process.

ITI has met with considerable success with its needs-driven approach, uncovering "new gold." Of the 64 technologies nominated in engagements with the first three labs, about 30% were technologies that had not been in any transfer or commercialization path, i.e., no patent activity was being pursued or invention disclosures filed. However, ITI concluded that the needs-driven approach should be brought to more labs, particularly smaller facilities and/or those within the DOD domain. Many labs are doing very little transfer work, and they need to understand where to start. ITI also found that much of the open literature on future technology needs is suspect and analytically weak and would benefit from a more systematic approach.

Another major limiting factor on the transfer of research-based technologies to industry is the lack of money for advanced preproprietary development. Lab-based technologies are not sufficiently hardened to show to potential commercializers, and labs often do not have the money to develop some of these underdeveloped technologies. Aside from the Small Business Innovation and Research Program (SBIR) few sources of funds can support the creation of engineering prototypes of promising inventions.

General Suggestions for Improving Technology Transfer

Clearly, if technology transfer is to succeed, the "culture gap" between the two sectors (industry and government labs) must be bridged through stepped-up interaction and communication, specifically between engineers and administrators from both sectors. Unless both sectors can reach a common definition and understanding of goals in a project, the innate differences in organizational cultures will persist to the detriment of effective technology transfer.

In the following sections, I make several specific suggestions as to how communication/information flows might be improved on all sides over the long term.

Government Labs Management

Each lab president must make a clear statement of action:

(1) Define a clear lab mission based on an assessment of industry activities and relevant priority technology lists and build on existing lab strengths and contacts. Provide highly specific definitions of technology projects.
(2) Concentrate on strategic or enabling technology sectors (printed wiring boards and multisequence machine tools are examples).
(3) Keep the industry-government agency (e.g., DOD, DOE) interface as simple as possible and avoid the "every constituency approach" (as in the CRADA model).
(4) Lab/government interaction is consistent with laboratory accountability, but not with detailed reviews, controls, and oversight from government department headquarters and from the operations offices. While the program needs general supervision at headquarters to give reasonable protection against embarrassment, this supervision should not burden or delay laboratory and industry interaction.
(5)   Insist on accountability in meeting goals based on these standards. Establish goals that are quantitative and not just general qualitative statements. Establish measurements against these goals.
(6) Rely on industry partners to make the evaluations of commercial adaptability, i.e., to make the "winners and losers" judgment; keep the

laboratories and government agency out of commercial policy, but encourage participating engineers to think in terms of the commercial application of technologies.

(7) Make sure that industry is directly involved in the conception, design, and management of any federal lab R&D programs devoted to generic technology before these programs are launched. Institute and encourage better/more information flows and interaction between labs and local industry.

(8) Institute personnel exchanges (both formal and informal). Use a multiplicity of specific technology sector workshops as a means of getting advice and "feedback" from industry.

(9) Develop benchmarks to gauge the degree of industrial involvement in the labs' research programs, the industrial relevance of their R&D in the near term, and the potential for applications from their work in generic technologies over the long term.

(10) Build relationships with other labs, e.g., the recent regional interlab strategic alliance involving three New Mexico labs. Expected benefits of such an alliance include a more effective use of government resources, a more rapid maturing of technologies, and a broader application of these technologies by government and industry.

(11) Institute and encourage extensive information flows/interaction between labs and local universities via personnel exchange and informal interaction.

(12) Actively seek out and arrange for interaction among intermediaries (such as those mentioned in the Bingaman legislation or the ITI project), businesses, and other labs, making use of the new regional manufacturing technology transfer centers.

(13) Institute and encourage information flows and interaction among lab personnel, i.e., divisions and departments. Institute interdepartmental personnel exchanges.

(14) Avoid confining technical staff to narrow and repetitive tasks.

(15) Incorporate TQM models and emphasize efficiency for technical staff. Run the lab like a business. Support technical staff participation in special activities such as those mentioned in Recommendation #11.

(16) Provide (creative) outlets for technical staff to increase personal motivation, communication, and leadership skills, as well as creativity.

Make the lab engineer's position more socially valued and rewarding by facilitating teaching sabbaticals or part-time positions with local school communities. (This would have the two-fold benefit of (1) getting more American schoolchildren interested and involved in science early on and (2) improving the social prestige and visibility of engineers.) Encourage attendance at seminars and provide a yearly awards structure.

Industry Management

(1) Provide more opportunities for interchange and building relationships between industry engineers and lab engineers. Factor lab engineers into project consulting processes, whether in a joint project, a participation by industry in a project funded by and for the government lab, or one in which there is a simple transfer of government lab technology.

   In joint projects, industry personnel should be involved at the earliest stages. Where there is a straightforward technology transfer, the originating scientist or team of scientists should—where feasible—remain involved in the development of commercial applications.

(2) Facilitate regular personnel exchanges, sabbatical leaves, and/or recon trips to labs.

(3) Actively seek out facilitators, if necessary, to link industry with labs conducting relevant research.

Engineers in Labs

(1) Learn to work *with* the "customer" in the planning phase of a project. Initiate and/or participate in excursions to local companies to understand the corporate culture and the imperative for rapid commercialization.

(2) Try to take a "synthesis" approach to problem solving rather than a strictly analytical approach. Get further training in other engineering disciplines, if possible, and try to take some coursework in business, as well.

(3) Make use of as many of the exchange programs offered by management as possible.

Academic Management

(1) As part of the training of the engineering student, focus on all the things that go into making a product more commercially relevant. Instruct engineering students to think in terms of feasibility and commercial adaptability of technological innovations.

(2) The engineering curriculum should include courses on design and systems considerations to augment the primary field of discipline.

   The engineering curriculum should stretch beyond mere analysis and a monodisciplinary approach. A multidisciplinary approach is needed, fostering greater ability to synthesize material and concepts in the design process. For example, mechanical and electronics engineers should be more informed about each other's work. A way to promote this interaction might be a senior-level class that would pose problems drawing on different engineering disciplines, as well as other relevant aspects such as environmental or economic considerations.

(3) In general, engineering instruction should impart a work ethic of teamwork, not isolation. The need for regular flows and exchange of information must be stressed. This is important because engineers show a natural tendency to value individualism above teamwork. Engineers should be encouraged to avoid "neat" compartmentalization of projects.

(4) Engineering instruction should stress the value of engineering to society.

Summary

A nation's ability to research and develop technology affects productivity, economic growth, and national security, as well as an overall perception of a nation's international power. Because the United States is faced with an ever-growing field of international players that are becoming increasingly skilled at researching and developing key technologies, America must step up its own efforts at technology development if it wishes to enjoy the economic and political position to which it has become accustomed over the past century.

One potentially effective means to strengthen our nation's technology base is through technology transfer from government labs to private industry. Previously banned from interacting through a host of legal, regulatory, and structural barriers, government labs and private industry are now being encouraged to cooperate more regularly. Although many of these barriers between industry and government have come down, real differences in the missions and constraints of each sector linger and conspire to make the lab-to-industry technology transfer process more difficult.

Greater communication between the two sectors and a shared understanding of the operating conditions and the goals of a given technology project are essential to successful technology transfer. Fundamental to this two-way communication is effective leadership from both lab and industry management, as well as a willingness by lab and industry engineers to cooperate and take a multidisciplinary systems approach to solving technology problems.

Notes

[1] *Gaining New Ground: Technology Priorities for America's Future*, Council on Competitiveness (Washington, DC: Council on Competitiveness, March 1991), 11.

[2] Source: Council on Competitiveness.

[3] *Gaining New Ground: Technology Priorities for America's Future*, Council on Competitiveness (Washington, DC: Council on Competitiveness, March 1991), 31.

[4] Ibid., 31.

[5] Ibid., 32.

[6] Source: Council on Competitiveness.

[7] *Gaining New Ground: Technology Priorities for America's Future*, Council on Competitiveness (Washington, DC: Council on Competitiveness, March 1991), 33.

[8] Ibid., 33.

[9] John R. Gawalt, Selected Data on Research and Development in Industry: 1989 (Washington, DC: National Science Foundation, Division of Science Resource Studies, February, 1991), Table SD-1.

[10] Source: Council on Competitiveness.

[11] Ibid.

[12] Productivity and the Economy, U.S. Department of Labor (Washington, DC: Bureau of Statistics, 1979).

[13] Ralph Landau, "Capital Investment: Key to Competitiveness and Growth," *The Brookings Review* (Summer,1990): 54.

[14] *Gaining New Ground,* 1.

[15] Sec. 3132. Findings and Purposes (National Competitiveness and Technology Transfer Act). Pub. L. 101-189 (Nov. 29, 1989).

[16] There is no agreement on the precise number of labs because of the definitional differences in what constitutes a national lab.

[17] *Gaining New Ground,* 47.

[18] Ibid.

[19] Ibid.

[20] Sec. 3132. Findings and Purposes (NCTTA) Pub L. 101-189 (Nov. 29, 1989).

[21] G.R. Bopp (Ed.), *Federal Lab Technology Transfer: Issues and Policies* (New York: Prager, 1988).

[22] Sec. 3132. Findings and Purposes (NCTTA), Pub L. 101-189 (Nov. 29, 1989).

[23] Kwai-Cheung Chan, Director, Program Evaluation in Physical Systems Area, U.S. General Accounting Office. Testimony before the Subcommittee on Technology and Competitiveness, Committee on Science,

Space and Technology.   U.S. House of Representatives (Washington, DC: May 30, 1991): 2-3.

[24] Jeff Bingham, U.S. Senator, "The Small Business-National Defense Laboratory Partnership Act of 1990," Federal Laboratory Consortium (FLC) for Technology Transfer NewsLink 7, No. 5 (May 1991): 1.

[25] Ibid.

[26] Ibid.

[27] Deborah L. Wince-Smith, Assistant Secretary for Technology Policy, U.S. Department of Commerce. Testimony before the Subcommittee on Technology and Competitiveness, Committee on Science, Space and Technology, U.S. House of Representatives (Washington, DC: May 30, 1991): 3-4.

[28] Ibid.

[29] George H. Kuper, President, Industrial Technology Institute (Ann Arbor, Michigan), Testimony delivered to the Technology and Competitiveness Subcommittee, Committee of Science, Space and Technology. U.S. House of Representatives (Washington, DC: May 30, 1991): 9-10.

[30] Wince-Smith, 7.

[31] Ibid.

[32] Diffusing Innovations: Implementing the Technology Transfer Act of 1986. (GAO) PEMD -91-23 (Washington, DC: General Accounting Office, May, 1991).

[33] Chan, 4.

[34] Ibid.

[35] Technology Transfer at the Department of Energy Defense Laboratories: A Report of the Office of Defense Programs, Department of Energy and the Lawrence Livermore National Laboratory, Institute for Technology and Strategic Research (ITSR) (Washington, DC: George Washington University, 1991).

[36] This information comes from testimony before Congress by George Kuper, President of ITI, 6. See earlier footnote for details.

7

The Labor Economics of Conversion: Prospects for Military-Dependent Engineers and Scientists

Joel Yudken and Ann Markusen

Abstract

The jobs of a substantial number of scientists and engineers (S&E) in the United States are at risk due to anticipated large-scale reductions in the military's procurement and R&D budgets over the next few years. It is projected that the U.S. will experience growing shortages of S&E personnel in a number of fields over the next decade. This could provide many opportunities for S&Es laid off due to military cutbacks. But, because of the concentrated nature of defense-dependent S&E employment and structural barriers to transferability of job skills and experience, ordinary market forces will not be sufficient to help S&Es make the transition from military- to civilian-oriented jobs.

A primary impetus in the growth of the S&E work force in the United States over the past four decades was the Cold War. The close link between S&E employment and military spending over this period is a product of the predominant role of the military in the U.S. R&D system and the high level of R&D intensity in military-industrial production. About two-thirds of the federal R&D budget and about one-third of all R&D spending in the nation is tied to national defense. The most highly defense-dependent industries—aerospace, electronics, and communications equipment—are also the most R&D-intensive and receive the largest share of federal R&D funds. Research and development is at the heart of most S&E activity. Between 35% and 40% of S&Es engage in R&D or management of R&D as their primary work. Many others work in administration, production, inspection, and sales related in some way to products of the R&D/design endeavor. The military's involvement in R&D must therefore be a crucial concern in evaluating the S&E labor market.

Around 15% to 20% of all employed scientists and engineers in the United States engage in defense-related work. This work force is very concentrated, however, in a relatively small number of industries (primarily durable goods manufacturing), firms, geographical regions, and occupations. The bulk of defense-related S&Es tend to work and live in just a few regions in the nation—particularly in the Pacific Coast, New England, and Middle Atlantic states.   Within certain occupations—electrical engineering, aeronautic and astronautical engineering, physics and astronomy, mathematics, and metallurgical and materials engineering—a minimum of 1/5 to 2/3 of the technical workers are defense-related.

Younger engineers who are not as overspecialized, with more up-to-date skills, and less "socialized" into the defense-industrial engineering culture, will fare better than older S&E workers who have been employed in defense jobs for long periods of time.   Studies show that S&E mobility between military and civilian sectors tends to be relatively high within firms but less so between firms and industries.   Mobility seems to be lowest within highly defense-dependent industries and highest in the less defense-related industries.   This suggests that defense engineers and scientists will have a harder time moving from their firms in the manufacturing-intensive military industries to jobs in the services-producing industries where the future demand for S&Es is expected to be highest.

Government-sponsored readjustment (income, retraining, and relocation) assistance for defense dependent S&Es will be necessary to help S&Es make the switch to jobs in the civilian economy, but will not be sufficient.   Facility conversion legislation to assist conversion efforts at the plant and community level will also be helpful, but still not adequate to absorb the large number of S&Es who will be displaced.   Engineers and scientists have played critical roles in prior conversion attempts initiated by both community/labor groups and corporations.   Although these efforts met with only mixed success, their experience suggests that, with some government assistance, conversion can be successful in some instances. But, even if successful, because of the great difference between the proportion of S&Es employed in military industry versus that in civilian industry, many S&Es still would not be able to find new employment.

What is needed is a combination of short-term adjustment assistance and longer-term demand creation policies for redirecting the nation's science and technology resources toward new national goals (environment, health, and economic stability).   Strong government R&D (and related) programs geared toward addressing vital economic, social, and environmental problems would be part of such a strategy.   The many new job opportunities for S&E workers and new markets for converted plants that such a program would create would make both occupational and facility conversion substantially more viable.

Further research is needed to (1) more accurately assess the projected pattern of S&E displacement expected as the Pentagon implements its defense preparedness plans over the coming decade;   (2) evaluate the potential supply and demand flows in the S&E labor force taking into consideration the Pentagon's pared-down defense budget and the economic competitiveness problems of U.S. manufacturing industries; and

(3) evaluate the potential impact of alternative national R&D polices tied to civilian economic, social, and environmental goals on the supply and demand of S&E workers in the coming decade.

Engineers and Defense Dependency

Introduction

Even as the Pentagon was raining its high-tech, high-precision missiles and explosives on Iraq, it dropped a small bombshell of a different sort out on New York's Long Island. The Navy, in a cost-cutting move that is expected to save nearly $16 billion by 1997, decided, in late February 1991, to cancel its contract with the Grumman Corporation to modernize the F-14 Tomcat fighter aircraft. The immediate impact of this cut will be the loss of several hundred jobs, with possibly more layoffs over the long term.[1] While Grumman and the communities where its plants reside have weathered worse shocks in the past, this event is the product of a much larger dynamic that began with the unwinding of the Cold War, which, despite the Gulf War, is expected to continue over the next few years: the gradual paring down and restructuring of the U.S. military strategic and conventional forces.

The Pentagon's aircraft spending, for example, will be only $16 billion in the 1991 fiscal year, down from $48 billion (adjusted for inflation) a year in 1985, the height of the Reagan arms buildup. As a result, the aerospace industry expected to lose by the end of 1991 one-third of the military aircraft work force it employed just four years ago, or 165,000 jobs, with more cuts to come. One aerospace CEO predicts that "In the future, I see one, two, even three companies either going out of business or downsizing significantly."[2] The defense industry has lived through major declines before. But as another industry executive observes, "The difference now is that we don't see what would bring [spending] up again."[3]

The military rollback of weapons programs and bases is a major threat to the economic security of numerous communities, businesses, and workers throughout the nation, especially in that string of regions—called the "gunbelt"[4]—that have become heavily dependent on the military's largesse over the past four decades. In response, political, corporate, labor, and public interest leaders alike—have shown great interest in adjusting, diversifying, and converting military economic activity to civilian economic activity, perhaps more than ever before.

Over the past years, much attention has been given to the problems of diversifying companies and local economies and to the pros and cons of converting military facilities, but many issues associated with ameliorating the impacts of military cutbacks are far from resolved. Too little attention has been given to the crucial problem of converting the work force most directly threatened by military cutbacks. Critics of conversion argue that economic adjustment assistance programs at the federal and state levels are adequate for handling most of the dislocation problems of military workers. Except for these minimal forms of economic assistance, they want to let the marketplace be the arbiter of defense workers' fates, as it has been for workers in civilian industries when plants have been closed. Others, most notably Seymour Melman and fellow conversion scholars, have argued extensively that defense workers represent a unique case, deserving and requiring special attention, with respect to making the transition from military to civilian production jobs.

On the special nature of military workers, at least, there seems to be fairly widespread agreement. Some of the unique characteristics of the defense-industry labor force, for example, have been noted by defense analyst Jacques Gansler. Military workers tend to be more highly skilled, older, and receive higher wages. Because of the unusually craft-like nature of military production, which is geared to turning out small numbers of very complex, customized, large-scale products, defense workers tend to have lower levels of production-rate "learning"[5] and productivity than their counterparts in civilian industry. Moreover, because of the high volatility of the military market, driven as it is by primarily political rather than economic forces, the "most significant characteristic of the defense labor market," Gansler argues, "is its extreme long-term instability."[6]

Yet, Gansler laments that given these "special labor considerations" in military industry, "it is shocking to realize how little attention is paid to this labor market—especially when one considers that 20%-30% of all U.S. scientists and engineers and 8%-10% of all factory workers belong in the defense categories."[7] The purpose of this chapter is to help fill this gap in understanding of the defense labor force, and in particular, to shed more light on one of the least understood sectors of that work force, the scientist and engineers.

It should be noted that because of the nature of the literature and data available, scientists and engineers are lumped together in most discussions about this work force. The roles of scientists and engineers, in fact, are often interchangeable, depending on the work setting. Many engineers engage in scientific research, and many scientists are involved in engineering tasks and projects. Whenever possible, relevant distinctions between these two classes of technical workers will be made. Otherwise, we refer to the scientist and engineering work force as S&Es, when such differentiation is not possible or meaningful.

The following discussion is driven by the general question of whether military-dependent S&Es threatened by cuts in military programs require special—and if so, what kinds of—consideration with respect to making the shift (i.e., "conversion") to civilian work. In addressing this question, we examine three specific questions:

- What is the extent and nature of the vulnerability of military-dependent S&Es to military cutbacks?
- What are the obstacles and possibilities of converting military-dependent S&Es to civilian jobs?
- What role can and should S&Es play in the processes of facility conversion and industrial, local, and regional diversification?

The S&E Labor Force in the United States

It is useful to evaluate these questions against the backdrop of the overall U.S. S&E work force. Several characteristics of the S&E work force are illustrated in Tables 1, 2, and 3. One thing that stands out right away is the great variability of the S&E work force. It is crucial to keep in mind the many differences between types and conditions of S&E workers when trying to assess the extent of military dependence of the S&E work force or design economic conversion and adjustment policies for them.

Based on National Science Foundation estimates, the total number of employed S&Es equaled an estimated 5.5 million in 1988.[8] About 52% or 2.8 million are engineers, which is a little larger than the scientists' work force, which numbered 2.6 million in 1988, representing 48% of the employed S&Es. By sector of employment, 68% of all employed S&Es work in business and industry, another 14% in academic institutions, 8% in the federal government, and the remainder in state and local governments, nonprofit, the military, or other employers. Scientists and engineers differ significantly in the breakdown by employment sector. Industry employs about 80% of the nation's engineers, but only 55% of the scientists. By contrast, academia accounts for 24% of scientists' employment, but only 4% of the engineers.

Primary Work Activity: In terms of primary work activity,[9] 27% of all S&Es engage in R&D. Surprisingly, more engineers (33%) than scientists, (21%) engage in R&D as their primary work. Nearly 90% of the R&D engineers, however, work on development projects, compared to 40% of the R&D scientists. Most R&D scientists are primarily concentrated either in basic or applied research activities. About 27% of S&Es, including 25% of the scientists and 30% of the engineers, were primarily engaged in management jobs, mostly non-R&D related. Production and inspection jobs occupied 17% of the engineering work force, but only 7% of the scientists. Conversely, scientists were far more heavily involved in reporting, statistical work, and computing jobs than engineers, occupying 17% versus 5% of the respective work forces. More than five times as many scientists as engineers were reported as engaged in teaching, while nearly twice as many engineers as scientists were consultants.

S&E Occupations: By far, computer specialists represent the largest level of employment of any S&E field, accounting for 13% of all

S&E jobs in 1988 (Table 1).   Within the science fields, social scientists, life scientists, and psychologists make up the largest blocs of employment, followed by physical, mathematical, and environmental scientists. Electrical/electronics engineers and mechanical engineers lead the engineering fields in terms of employment, each with nearly 12% of the total S&E labor force.   Civil engineers are the next largest group, followed by industrial, aeronautical/astronautical, chemical, and materials engineers.

Although computer specialists are lumped in with the scientific fields in most S&E data sets, in many ways it constitutes a unique—and increasingly important—category of technical professional, which spans a diverse set of subfields and types of activity within the S&E occupations.   At one end of the spectrum are the computer scientists, most of whom primarily engage in R&D activities—roughly one-fifth of the computer specialists— which encompasses a conglomeration of specialties ranging from applied mathematicians to software scientists to microsystems engineers.   The broad subfield of artificial intelligence even includes cognitive scientists among its ranks.   At the other end of the field are the systems analysts and computer specialists whose primary activities include computer applications, statistical analysis and information retrieval.   Nearly half of all computer specialists fall into this latter group.   In fact, computer specialists account for about 60% of all S&E workers and more than three-quarters of all scientists whose primary work activities are reporting, statistical analysis, and computing.[10]

Industrial Sectors: Within private industry the pattern of employment of S&Es, particularly that of engineers, is significantly different from that of the total industrial labor force.   As Table 3 shows, about two-thirds of engineers in industry are employed in goods-producing industries, mostly concentrated in the durable goods manufacturing sector.   Only one-third of the engineers work in the services-producing industries, the bulk of which are employed in the business and related services, which include R&D, engineering, and professional services to the manufacturing sector. By contrast, two-thirds of the overall industrial labor force works in service industries, primarily in trade, followed by business and related services. Durable goods producing industries are only the third largest sector of employment for all occupations.

At first glance, scientist employment in industry appears to follow the total work force pattern, with over 60% employed in service industries, except that the largest segment of scientists are in business and related services, followed by durable goods and financial services.   However, computer specialists account for 60% of all scientists employed in service industries.   Nearly 30% of the computer specialists are employed in the business and related services industry, and another 20% in financial services.   If this category is deducted from the scientist data, then only a little more than half of all industrial scientists are employed in services, while a little less than half work in goods-producing industries.

Table 1
S&Es Employed by Field, 1988[11]

FIELD	TOTAL S&E EMPLOYED (000s)	% OF TOTAL S&ES
Total, all fields	**5475**	**100.0**
Total Scientists	**2625**	**47.9**
Physical scientists	311	5.7
Chemists	196	3.6
Physicists/astronomers	78	1.4
Other physical scientists	37	0.7
Mathematical scientists	167	3.1
Mathematicians	141	2.6
Statisticians	26	0.5
Computer specialists	710	13.0
Environmental scientists	113	2.1
Earth scientists	94	1.7
Oceanographers	5	0.1
Atmospheric scientists	14	0.3
Life scientists	460	8.4
Biological scientists	299	5.5
Agricultural scientists	124	2.3
Medical scientists	38	0.7
Psychologists	334	6.1
Social scientists	529	9.7
Economists	218	4.0
Sociologists/anthropol.	94	1.7
Other social scientists	217	4.0

Table 1 Cont.

FIELD	TOTAL S&E EMPLOYED (000s)	% OF TOTAL S&ES
Total Engineers	**2850**	**52.1**
Aero-/astronautical	119	2.2
Chemical	150	2.7
Civil	338	6.2
Electrical/electronics	639	11.7
Industrial	172	3.1
Materials	66	1.2
Mechanical	649	11.9
Mining	21	0.4
Nuclear	28	0.5
Petroleum	39	0.7
Other engineers	630	11.5

Table 2

Selected Characteristics of S&E Work Force

SECTOR OF EMPLOYMENT-1988[12]	PERCENT OF ALL EMPLOYED		
	S&ES	SCIENTISTS	ENGINEERS
Total	**100**	**100**	**100**
Business and Industry	68	55	80
Industry	62	48	75
Self-employed	6	6	5
Educational Institutions	14	24	4
4-year	10	17	4
Other	3	6	1
Nonprofit Organizations	4	6	2
Federal Government	8	8	7
Military	1	0	1
*Other/no report***	7	7	6

** Including other government

Table 2 Cont.

PRIMARY WORK ACTIVITY-1988[13]

Total	100	100	100
Research & *Development*	27	21	33
Basic research	3	5	1
Applied research	5	7	4
Development	19	8	29
Management/ *administration*	27	25	30
Management of R&D	8	7	10
Other than R&D	19	18	20
Teaching	8	14	2
Consulting	5	4	7
Production/inspection	12	7	17
Reporting, Statistics, *Computing*	11	17	5
Other/no report**	9	13	6

** Including sales and professional services

HIGHEST DEGREE ATTAINED-1986[14]

Total	100	100	100
Doctorate degree	13	23	4
Masters degree	24	24	23
Bachelors degree	57	52	60
Other	6	1	13

GENDER AND ETHNIC GROUP-1988 (EST.)[15]

Total	100	100	100
Men	84	71	95
Women	16	29	5
White	90	90	91
Black	3	4	2
Asian	5	5	6
Other	2	2	2

Table 3

Scientist and Engineer Employment by Industry, 1988[16]

A. PERCENT OF ALL OCCUPATIONS IN INDUSTRY

| | | PERCENT OF ALL OCCUPATIONS | | |
| | TOTAL ALL | ALL | ALL | ALL |
INDUSTRY	OCCUPATIONS [17]	S&ES	SCIENTISTS	ENGINEERS
All Private Industry	**77,102**	**2.4**	**0.8**	**1.7**
Goods-Producing:	25,250	4.2	0.9	3.3
Durable goods	11,437	7.0	1.0	6.0
Nondurable goods	7967	2.3	1.1	1.2
Mining	721	6.7	3.1	3.7
Construction	5125	0.5	[18]	0.5
Services Producing:	51,852	1.6	0.7	0.9
Communications/ transportation/utilities	5548	2.0	0.6	1.4
Trade	25,138	0.3	0.1	0.2
Financial services	6676	1.8	1.6	0.2
Business & related services	14,490	3.4	1.3	2.1

* Less than 1/2 of 1%.
**Figures in thousands.

B. PERCENT OF ALL PRIVATE INDUSTRY

| | | PERCENT OF ALL PRIVATE INDUSTRY | | |
| | TOTAL ALL | ALL | ALL | ALL |
INDUSTRY	OCCUPATIONS	S&ES	SCIENTISTS	ENGINEERS
All Private Industry	**100**	**100**	**100**	**100**
Goods-Producing:	32.7	56.7	38.6	65.0
Durable goods	14.8	42.9	19.6	53.6
Nondurable goods	10.3	9.9	14.9	7.6
Mining	0.9	2.6	3.8	2.1
Construction	6.6	1.3	0.2	1.9
Services Producing:	67.3	43.3	61.6	35.0
Communications/ transportation/utilities	7.2	6.0	5.5	6.2
Trade	32.6	4.1	6.2	3.2
Financial services	8.7	6.6	18.4	1.2
Business & related services	18.8	26.6	31.4	24.3

Some totals do not add up due to round-off.

As we will see later, the pattern of S&E employment within industry shown in Table 3a, is linked to the heavy R&D emphasis of the military industries, which are major employers in the durable goods manufacturing sector. For example, engineers comprise 6% of the durable goods sector's labor force, but under 2% of that in all private industry and less than 1% in the service sector. Only in mining, which includes the petroleum industry, do we see relatively high S&E employment as a share of the overall work force. Business and related services also employ a larger proportion of S&Es compared to the total industry pattern.

Gender and Race Composition: The S&E work force differs substantially from the total labor force in its gender and race composition as well. Although progress has been made over the past decade, women and minority ethnic groups remain underrepresented in the S&E labor force. Science and engineering remain predominantly white male occupational preserves. This reflects the generally low involvement of these groups in the higher-skilled occupations in the overall work population. Although women make up over 40% of the employed civilian workforce, they were only 16% of the S&E workforce in 1988 (estimated).[19] They are much better represented, however, in the science professions, accounting for 29% in 1988 of the employed scientists, than in the engineering work force, of which only 5% are women.

Similarly, although blacks represent about 10% of total U.S. employment, and almost 7% of professional and related occupations, they are under 3% of the S&E work force. Again, blacks do better in science, accounting for 4% of those occupations compared to 2% of the engineering jobs. Asians, in contrast, who represent 2% of the total work force and 3% of the professional and related occupations, occupy 5% of the S&E jobs.[20] The lack of representation of women and racial minorities in the S&E work force is an important issue that deserves attention in any discussion about economic conversion strategies for scientists and engineers. Military industries are not necessarily any worse than civilian employers with respect to recruiting S&Es from these underemployed groups. However, the lock that these industries have had on certain S&E occupational groups may have been a factor limiting new opportunities in the civilian sector that might have more readily attracted minorities and women into the S&E professions.[21]

Doctoral S&Es: Women and ethnic minorities do not fare any better in the employed S&E population with doctorates.[22] People of Asian descent are again an exception, making up 7% of the scientists and 19% of the engineers with doctoral degrees. Although representing only 13% of employed S&Es—23% of scientists, but only 4% of engineers—doctoral S&Es are an especially important part of the technical professional work force.[23] They are the core workers in the nation's research and innovation system—over a third of this group engage in R&D as their primary activity, mostly in basic and applied research—and in the educational system that trains new generations of S&Es—over a quarter consider teaching their primary work activity.

About half the doctoral S&Es work in academia and a little under a third work in industry.[24] Among S&E fields, the life sciences by far account for the largest share—over a quarter—of employed doctoral S&Es (see Table 4). The physical, social, and psychological sciences have the next largest doctoral populations. The mathematical, computer, and environmental sciences each account for only 4% of total S&E doctoral employment. Electrical and electronics engineering employ the largest number of doctorates among the engineering fields but account for only 3% of all S&E doctoral employment. This is roughly twice the share of the next largest fields, chemical, civil, and mechanical engineering. The low showing of engineering in the doctoral population is partially offset by the large number of engineers with professional engineer degrees (included in the Other category in Table 3). It also reflects the fact that engineering graduates are less likely than scientists to require Ph.D.s in order to advance in their careers.

Special attention must be given to the problem of conversion of the doctoral S&E work force, between 40% and 50% of whom receive some form of federal support,[25] a large portion from military agencies, to carry out their work. Cuts in military programs will affect this part of the work force very differently from the average production S&E worker employed in industry. These distinctions suggest that different approaches may be required for each category of S&E worker to address the problems of making the adjustment from military to civilian work.

Rise of the U.S. S&E Labor Force

The modern professions of scientist and engineer evolved alongside the manufacturing system given birth to by the Industrial Revolution. The engineering profession barely existed in the early 19th century. According to Braverman, "it has been estimated that there were no more than some 30 engineers or quasi-engineers in the United States in 1816." An 1850 census showed the existence of about 2000 civil engineers at that time. With industrialization, though, both the number and categories of engineering increased. Between 1880 and 1920, the number of engineers grew from 7000 to 136,000, now including mining, metallurgical, mechanical, electrical, chemical, and civil engineers.[26] Driven by Taylorism, scientific management, and the increasing reliance of modern production practice on science and technology—not to mention the technical requirements of modern industrial warfare—the S&E professions grew at a spectacular rate during the middle part of the 20th century, far greater than that of the overall work force. By 1930 the number of engineers jumped to 217,000, and by 1950 it increased to 535,000. Between 1950 and 1966 engineering employment rose 80% to about 1 million, a figure that has since escalated by a factor of almost 3 to the current level of 2.8 million employed engineers. The number of scientists also mushroomed, growing by 930% between 1930 and 1964, from 46,000 to nearly a half a million.[27]

According to Daniel Bell, the rise of the professional classes, including S&Es, which was part of an equally dramatic increase of all white collar workers, including administrators and clerical workers, is evidence of the emergence of a new post-industrial society. Indeed, professional, technical, and kindred workers, which was only the sixth largest category of workers in 1940, rose to the fourth largest major occupational group by 1960, and to the number two spot by 1980.[28] It should be recalled, however, that while S&Es are indeed "information" producers and users, the majority are employed in goods manufacturing industries. And among the biggest employers in this sector are military industries.

There is little question that both World War II and the Cold War gave tremendous impetus to S&E employment in the United States over the last four decades. Bell notes that the main reason for this growth in the 1950s was "the expansion in this period of the science-based industries such as electronics, space, missiles, scientific instruments, nuclear energy, and computer technology. Not coincidentally, these same industries were the ones most heavily tied to military R&D and production. Similarly, the jump in S&E jobs after 1960 was fueled partially by post-Sputnik and space-program-driven government policies to recruit "baby boomers" into the S&E professions, and in the 1980s by the massive military buildup under President Reagan.

S&E Growth and Military R&D

The close link between S&E employment and military spending over the postwar era is the product of two interrelated factors: the predominant role of the military in the U.S. R&D system since W.W.II, and the high level of R&D intensity in military-industrial production. Research and development is at the heart of most S&E activity. Between 35% and 40% of S&Es engage in R&D or management of R&D as their primary work (Table 2). Many others work in administrative, production, inspection, and sales related in some way to products of the R&D/design endeavor. The military's involvement in R&D must therefore be a crucial concern in evaluating the S&E labor market.

The modern research and development system in the U.S. evolved out of W.W.II. The Manhattan Project, which built the first atomic bomb, was a prototype and model for the government-academic-industrial collaboration on R&D projects, which has characterized much of the U.S.

Table 4

Doctoral S&Es Employed by Field, 1987[29]

FIELD	TOTAL S&E EMPLOYED	% OF TOTAL S&ES
Total, all fields	**419,118**	**100.0**
Total Scientists	351,350	83.8
Physical scientists	68,647	16.4
Chemists	44,136	10.5
Physicists/Astronomers	24,511	5.8
Mathematical scientists	16,699	4.0
Mathematicians	13,878	3.3
Statisticians	2821	0.7
Computer specialists	18,571	4.4
Environmental scientists	17,811	4.2
Earth scientists	13,577	3.2
Oceanographers	2037	0.5
Atmospheric scientists	2197	0.5
Life scientists	107,378	25.6
Biological scientists	61,985	14.8
Agricultural scientists	15,796	3.8
Medical scientists	29,597	7.1
Psychologists	56,378	13.5
Social scientists	65,866	15.7
Economists	17,837	4.3
Sociologists/ anthropologists	12,933	3.1
Other social scientists	35,096	8.4
Total Engineers	67,768	16.2
Aero-/astronautical	5005	1.2
Chemical	6923	1.7
Civil	6479	1.5
Electrical/electronics	12,601	3.0
Mechanical	6711	1.6
Nuclear	2151	0.5
Other engineers	27,898	6.7

science and technology enterprise ever since.   Up to 1960 the military provided 80% of total federal support to R&D.   As civilian agencies, such as the National Aeronautics and Space Administration (NASA), National Science Foundation (NSF), and National Institutes of Health (NIH), came into their own during the 1960s and 1970s, federal civilian R&D funding achieved a rough parity with military R&D spending.   Military R&D support was also depressed as a consequence of the Vietnam War and its aftermath.

The Reagan buildup of the early 1980s reversed this trend.   Defense R&D jumped from 49% ($13.8 billion) of the total federal R&D budget in fiscal 1979 to 69% ($36.9 billion) in fiscal 1986.   This was equivalent to a 9% per year real growth rate while real civilian R&D spending declined by about 4% a year.[30]   Due to budget pressures and the winding down of the Cold War military R&D growth has slowed down since 1986.   The 1990 estimated defense R&D budget ($39.9 billion) was only 62.5% of total estimated federal R&D expenditures ($63.8 billion).   Nevertheless, as federal R&D spending accounts for half of all R&D spent in the U.S., roughly one-third of the nation's R&D activity is military-related.

Although about a quarter of Department of Defense (DoD) R&D[31] is done by its own labs, industry is the military's principal R&D performer; it gets about two-thirds of all Pentagon R&D dollars. A distant third and fourth place, are the Federally Funded Research and Development Centers (FFRDCs)—government funded research facilities operated by private industry, academic institutions, or nonprofit corporations, which includes the vast DOE national laboratory system—which receive about 4% of DoD R&D funds, and universities and colleges which receive 3% of this largesse.[32]

Over 90% of the military R&D budget falls into the development category, mainly large weapons systems programs, of which the Strategic Defense Initiative, weighing in at $3 billion to $5 billion a year, is among the largest. Industry performs about 70% of this work, and about 47% of DoD's applied research, which accounts for 6%-7% of the DoD R&D budget. Only 2%-3% of this budget is devoted to basic research, of which 60% is performed by universities and colleges.[33] DoD has played an increasingly smaller role in basic research over the years, dropping from 14% in 1969 to 9% in 1990 of total federal support for basic research.[34]

Industry and Military R&D: The military's prominent role in industrial R&D is of special concern. We calculate that approximately one-third of *all* industrial R&D is directly or indirectly defense-related.[35] Defense industry is unusually manufacturing intensive. About 40% of the military dollar goes to the manufacturing sector, either as orders or as research contracts for future products, compared with manufacturing's 20% share of the entire economy. David Henry's work at the Department of Commerce reveals an elite set of industries tightly linked to the Department of Defense market.[36] Topping the list are the key members of the aerospace (aircraft and parts, guided missiles, and spacecraft) sector plus shipbuilding, ordnance, and tanks. These industries all had 40% or more of their capacity committed to the Department of Defense by 1985. The biggest two clusters are aircraft and missiles, which with their engine and parts suppliers landed $22.6 billion in DoD sales in 1985, and communications equipment, which sold $15.7 billion to the same customer. If we include defense-related NASA purchases, the Department of Energy's orders for nuclear warheads, and foreign military sales, the size of these industries' military business would increase significantly.[37] These are *defense-dependent* industries.

Another set of industries, which we call *defense-related* , include high-tech supplier sectors like instruments and electronics as well as the more mature machine tools industry, which all had between 20% and 39% of their market dedicated to the military at the height of the 1980s buildup. With the exception of electronic components, their total receipts were small relative to those reaped by the defense-dependent industries.[38]

The lion's share of the military budget, however, goes to just three industries, aerospace, communications equipment, and electronics components, comprising what we refer to as the ACE complex, which in 1990 received 79% of all DoD procurement outlays. These are also the most R&D-intensive of both defense and all manufacturing industries. Over the decades, they have commanded the best of the nation's scientific and engineering expertise and reaped the bulk of public R&D subsidies. By the 1960s, the aircraft and missile companies alone employed more scientists and engineers on R&D work than the combined total of the chemical, drug, petroleum, motor vehicle, rubber, and machinery industries.[39] In funding, aerospace, communications, and electronics garnered $23 billion in R&D support from the federal government in 1989. By contrast, the steel industry received a minuscule $21 *million* (see Table 5). In the same year, the ACE complex received 81% of all federal funding for manufacturing R&D. By the same token, the aerospace industry depended on the federal government for 82% of all its R&D funds, and the communications and electronics industries, together, for 43% of their R&D monies. While the aerospace industry spent only slightly more of its own R&D funds as a percentage of sales than did the average manufacturing firm (3.7% compared with 3.2%), its total R&D funds as a percentage of sales were a stunning 15% compared with less than 5% for all manufacturing, the gap almost entirely due to military subsidy.[40]

R&D and S&E Demand: The tremendous rise in military R&D spending during the buildup of the 1980s, tied to new generations of large-scale strategic (and conventional) weapons produced by these industries, was a major stimulus in the rising demand for S&Es. The decade 1976-1986, covering the buildup period, witnessed an expansion of 100% in the number of scientists and 90% in the number of the engineers primarily engaged in R&D.[41] During the 1980s buildup, professional, technical, and service jobs increased by 82% compared to only 57% for blue collar precision and craft workers and 48% for less skilled operatives and fabricators.[42]

Industry accounted for over 70% of the 1.8 million person growth in S&E employment in the U.S. In the goods-producing industries, S&E jobs grew at an average rate of 2.7% per year between 1980 and 1988. Within the durable goods manufacturing sector, although total employment declined, S&E growth was nearly 8% per year.[43] This corresponds to the overall employment trends in manufacturing, in which the massive growth of an S&E-intensive defense industry labor force was only partly able to offset the rapid decline in the civilian manufacturing's predominantly blue collar labor force between 1980 and 1985.

During the 1980s the defense-related population grew enormously, reaching a high of 6.7 million by the late 1980s (including armed service personnel). About 1.2 million new jobs were created, an increase of 22%.[44] Of this, private sector was the biggest gainer with 993,000 new jobs, an increase of 46%. The buildup was a major factor in overall job growth, as well. By conservative estimates defense demand accounted for at least 17% of total job growth in the U.S. Defense demand accounted for 30% of the growth of goods producing industries in the first half of the decade and 54% of all defense-related private jobs created from 1980-1985 were in manufacturing. Nevertheless, while defense spending created 600,000 new manufacturing jobs in this period—the biggest chunk, 400,000, were in five major defense hardware industries, aircraft, shipbuilding, ordnance, missiles, communications equipment—total manufacturing employment fell by 1 million. For every job added in defense manufacturing, three were destroyed in the nondefense plants. In durable goods manufacturing, total jobs fell by 680,000 despite an increase of defense durable jobs of 580,000.[45] In short, we see a remaking of the manufacturing face of the U.S. in which employment growth in this sector became increasingly dependent on military sales. This is an important consideration in gauging the conversion potential of all defense-related manufacturing workers.[46]

The rapid growth of S&E employment in the manufacturing divisions during this period, in the face of an overall decline in jobs, can be attributed, therefore, in large part to escalating defense expenditures on R&D and new procurement, though it is important to recognize such factors as technological competition, high-technology capital investment, and civilian R&D expenditures, as other generators of S&E employment.[47] It is not surprising that by the end of this growth period the major S&E employers in the durable goods industries—which is the largest provider of S&E jobs in the good-producing sector—in 1988 were aerospace, with 24% of S&E jobs, communications equipment with 15%, office and computing with 12%, and electronic parts with 12%.[48] Certain S&E subfields also benefited more than others as a result of high defense expenditures. Electrical/electronics experienced the greatest growth of any engineering field. Also leading the growth in doctorate-holder employment was aeronautical/astronautics, electrical and electronics, and nuclear engineering, that a National Science Foundation observes, "likely has been sustained by defense-related requirements."[49]

For the sake of balance, it should be noted that another major growth area for S&E employment was the services-producing sector, with the largest gains being made in the business and related services and financial services industry. This reflected the general economic and employment growth in the economy over the past few decades favoring an expanding service sector, while manufacturing shrunk both in output and jobs. The revolution in information technologies—itself a by-product of long-term military R&D investment in the computing field—and the concurrent strong demand for information/data services created increasing employment opportunities for S&E workers in computer and data-processing services. Hence the spectacular rise of the computer specialist work force since 1976, along with growing numbers of new jobs for mathematical and social scientists.[50]

The Defense-Dependent S&E Labor Force

The link between S&E employment and military spending has long been a central issue in the debate about defense-dependency. "Depletionist" scholars, such as Lloyd J. Dumas at the University of Texas, Dallas, and physicist Warren Davis, who hold that the military industrial system drains off a disproportionately large share of the nation's top scientific and technical talent, thereby undermining the nation's economic health, tend to place the number of engineers and scientists engaged in defense-oriented activity at a lower bound of 30%. Occasionally, it has even been argued that as much as 50%-60% are so employed.[51] Jacques Gansler, as seen earlier, places the figure closer to between 20%-30%, though the basis for this estimate is not given.[52]

At the lower end, a National Science Foundation study based on "two state-of-the-art simulation models," estimated that "By 1987, roughly 4% of scientists, 15% of engineers, and 7% technicians are projected to be employed in defense activities, a slight increase over the 3%, 12%, and 6%, respectively, employed in such activities in 1982."[53] Similarly, 1982 data derived from the Pentagon's Defense Economics Impact Modeling System (DEIMS) projects that only about 14% of the nation's engineers and scientists working in industry should be included in so-called defense induced employment.[54]

Our own rough estimates, which are based on comparing two separate surveys, place the number of S&Es engaged in defense-related work at between 15% and 20%. However, if we make certain qualifying assumptions this figure goes higher, as will be shown. For reasons similar to those articulated by Dumas and Davis, we believe also that this estimate represents a lower bound. For instance, the surveys upon which the estimates are based may not have included S&Es working on military-related projects funded by ostensibly nonmilitary sources such as NASA, the so-called civilian space agency, or even the DOE.[55] Approximately half of the scheduled space shuttle missions are military-related, for example. A second source of undercounting, suggested by Warren Davis, is "that many companies that appear to be wholly commercial in nature and that do not... require security clearance for their employees, actually make significant contributions to defense technology." Digital Equipment Corporation (DEC), a leading computer manufacturer not known for its defense contracting, nevertheless sells computers, Davis notes, under government contract for "direct inclusion in defense applications, or by other companies for sole or shared use in fulfilling defense contracts."[56] That is, many S&Es may owe their livelihood, at least in part, to defense-related sales, without knowing it. It is not clear the extent to which this indirect or other "induced" defense-related employment is incorporated into the surveys and models upon which the defense-dependency ratios are calculated.

All these varied attempts to calculate defense dependency illustrate the many shortcomings and pitfalls in the measurement tools used. The rich, complex texture of the S&E labor force profiled earlier suggests the need for caution in generalizing about defense dependency based on aggregated data. It is difficult to get away from these problems even when trying to measure less aggregated categories. The computer specialists field is a case in point. Computer scientists employed in R&D and in some, but not all, durable goods industries are particularly reliant on military spending for their livelihoods, while the jobs of computer specialists employed in service-sector industries doing business-related computer applications are much less tied to military dollars. Yet both subfields are lumped together in the data estimates about defense dependency.

In any event, it is at the less aggregated levels that a more interesting picture arises concerning defense dependency. Despite the relatively low proportion of defense-reliant S&Es in the overall work force, they appear in much larger concentrations in certain industries, firms and institutions, occupations, and even regions. These more localized instances of defense dependency are important, as they represent different conditions affecting the prospects for converting the S&E work force.

S&E Concentrations in Defense Industries and Firms: Defense-dependent and related industries tend to employ S&Es in greater proportion than civilian industries, or even the civilian components of these same industries. As the Cold War evolved, the white collar work force, particularly S&E occupations, in aerospace industry began to rise faster than blue collar labor.[57] From 1954-1959, for example, electronics engineers increased from 16% to 25%.[58] By 1970, engineers comprised only 1.2% of all employees in nonagricultural enterprises in the United States. But in the military electronics industry they represented, along with technicians, one-

Table 5

Federal versus Industry Funding for Selected Industries, 1989[59]

INDUSTRY	FEDERAL FUNDS	INDUSTRIAL FUNDS	TOTAL R&D FUNDS	FED % OF TOTAL R&D	% OF FED FUNDS
	$ million	$ million	$ million		
Aerospace*	15,647	3511	19,157	81.7	53.5
Electronics and communication*	7928	10,618	18,546	42.7	27.1
Rubber products	313	930	1243	25.2	1.1
Autos, trucks, RR(inc.tanks)*	1982	9431	11,413	17.4	6.8
Scientific instruments	991	5531	6522	15.2	3.4
Machinery (inc. computers)*	1669	10,457	12,126	13.8	5.7
Fabricated metals	73	732	805	9.1	0.2
Iron and steel	21	601	622	3.4	0.1
Chemicals	381	11,134	11,515	3.3	1.3
Petroleum products	21	2068	2089	1.0	0.1
Food & beverage	0	1172	1172	0.0	0.0
Paper/pulp	0	1009	1009	0.0	0.0
Textiles	0	176	176	0.0	0.0
Manufacturing	**29,233**	**59,648**	**88,871**	**32.9**	**100.0**

* Military R&D-intensive industries

third of all people employed. Yet, in the civilian-oriented part of the electronics industry, the engineering-technical share of employment has been about 11%.[60]

By 1980, the gap between military and nonmilitary manufacturing was quite striking (Table 6). Industries with high rates of defense dependency and/or large defense receipts were heavily represented among those with disproportionately large shares of their work force in S&E occupations. The missiles industry is the most S&E-intensive, with more than 40% of its work force in these occupations, compared with less than 6% for manufacturing as a whole. Computing, scientific instruments, and communications equipment follow, each with more than three times the manufacturing average.

The high concentration of S&Es in defense industries extends to comparably high levels of S&E employment among the major military contractors. These firms usually receive substantial R&D contracts for major weapons systems, which usually precede large procurement orders.

As Table 7 shows, only four firms among the top 25 DoD contractors in 1987 garnered nearly about a third of all DoD RDT&E prime contracts, and twenty received nearly 60%. Landing a major R&D contract can result in a big chunk of new S&E jobs. In a single week in 1984, Lockheed's Space and Missile Division in Silicon Valley announced that it would hire 2600 new people to work on the Trident ballistic missile and the MILSTAR communications satellite. Of these, 25% would be electrical engineers and 16% computer scientists.[61]

On the other hand, some of the more mature defense industries, such as those producing ships and tanks, remain primarily blue collar intensive. In private shipyards, production workers may account for as much as 80% of the work force. In contrast, a large defense electronics firm like Hughes has 55% of its employees in technical occupations. Over 40% of Hughes' work force hold engineering degrees.[62] In the service end of the industry, where computer programmers model Star Wars or technical assistance firms aid the installation and operation of equipment, the S&E component can be even higher. At TRW's Washington, D. C., area systems house, two-thirds of the 1500 employees are engineers, and virtually all those employed at Ford Aerospace's Star Wars operations in Colorado Springs are white and pink collar workers.

Defense Dependency of S&E Occupations: Occupations, like industries, exhibit skewed ratios of military dependency, according to recent work by the Bureau of Labor Statistics (BLS) and work by Ann Markusen and Scott Campbell using the National Science Foundation's 1980 post-census Survey of Scientists and Engineers (SSE).[63,64] Comparing the BLS results for scientists and engineers with estimates generated from the self-reporting of these groups in the SSE, Markusen and Campbell found that while the order of occupations for both data sets are similar, there were systematic differences, especially at the "high" end (Table 8).

In both cases, aeronautical/astronautical, electrical/electronics, mechanical, metallurgical/materials, nuclear, and industrial engineers were at the top of the list of the most defense-dependent S&E occupations. But, aeronautical and astronautical engineers surveyed in the SSE reported a rate of dependency on DoD/NASA funding of 69%, versus the 37% estimated by the BLS method. The SSE defense share estimates exceed by four times those of the BLS for mathematicians, physicists, and other physical scientists, and are more than twice as high for electrical engineers and computer scientists. In only five of the twenty-eight occupations do the BLS figures exceed those of the SSE—industrial, chemical, petroleum and mining engineers, and agricultural scientists—and here the differences are relatively small, except for petroleum engineers.

The BLS estimates of defense dependency do not include NASA, DOE, or other non-DoD military projects or military projects for foreign governments. Adjusting for these omissions could add 15% to the shares generated by the BLS. Furthermore, the BLS makes assumptions in its models that may introduce significant bias. The most serious problem is the BLS assumption in its input-output model that defense-related sectors employ occupational groups in the same proportion as civilian components in the same industry. But, as we saw before, most defense firms tend to be much more S&E intensive than their civilian counterparts. This assumption, therefore, may bias the BLS estimates downward.[65]

Matching the SSE data with NSF S&E employment data for 1986 enables us to roughly estimate the overall defense dependency of the employed S&E population. We find that at least 15% of all employed S&Es engage in defense-related work, including 20% of all engineers but only 9% of all scientists. Certain fields in the sciences, particularly physicists/astronomers, other physical scientists, oceanographers, chemists, mathematicians, and computer specialists, have dependency levels much higher than that of all scientists. If we include only the physical, mathematical, environmental, and computer scientists in our calculations—i.e., deduct life, social, and psychological scientists—then the defense-dependent proportion of scientists rises to 15% and of all S&Es to 18%-19%.[66] Even these figures are low estimates, because additional defense-related S&Es are supported by Department of Energy work, by work associated with arms sales abroad, and by nonreimbursed R&D, which companies do in anticipation of future Pentagon sales.

Table 6

Concentrations of Scientists and Engineers in Manufacturing, 1980[67]

RANK	INDUSTRY	ENGINEERS AND SCIENTISTS % OF TOTAL EMPLOYMENT
1	Space vehicles & guided missiles	40.9
2	Office computing machines	26.6
3	Engineering, laboratory & scientific instruments	25.7
4	Communications equipment	21.3
5	Optical instruments & lenses	18.7
6	Aircraft & parts	18.0
7	Industrial organic chemicals	14.5
8	Measuring & controlling instruments	13.9
9	Electronic components & assembly	12.7
10	Petroleum refining	11.8
11	Engines & turbines	10.2
12	Industrial inorganic chemicals	9.5
13	Plastics & synthetic resins	9.4
14	Ordnance	9.4
15	Drugs	8.9
	Total manufacturing	**5.5**

Table 7

*RDT&E Awards to Top Defense Contractors*1987*[68]*

COMPANY	1987	% OF TOTAL
McDonnell Douglas Corp.	1669	7.7
Lockheed Aircraft Corp.	1631	7.5
Martin Marietta Corp.	1608	7.4
Boeing Corp.	1224	5.6
Grumman Corp.	902	4.1
General Electric Co.	784	3.6
Raytheon Co.	594	2.7
TRW, Inc.	557	2.6
Rockwell International Corp.	494	2.3
IBM Corp.	471	2.2
General Motors	395	1.8
Westinghouse Electric Corp.	385	1.8
Honeywell, Inc.	341	1.6
General Dynamics Corp.	334	1.5
United Technologies Corp.	282	1.3
LTV Corp.	253	1.2
Unisys Corp.	176	0.8
Ford Motor Corp.	172	0.8
ITT Corp.	163	0.8
The Singer Co.	147	0.7
US total RDT&E awards	21,809	
Share of top 5 in total	32.3%	
Share of top 10 in total	45.7%	
Share of top 20 in total	58.0%	

* Includes only those firms in DoD's list of top 25 prime contractors

Table 8

BLS and SSE Estimates of Defense Dependency by Occupation, 1986[69]

OCCUPATION	PERCENT OF OCCUPATION	
	BLS	SSE
Aeronautical and astronautical engineers	36.6	68.7
Oceanographers	NA	50.0
Physicists, astronomers	7.4	33.9
Electrical and electronics engineers	14.6	31.6
Metallurgical and materials engineers	16.5	24.2
Mathematicians	4.8	20.3
Physical scientists	2.8	19.9
Mechanical engineers	12.0	17.6
Other engineers	14.7	17.2
Nuclear engineers	10.0	15.8
Industrial engineers	14.4	13.4
Computer specialists	5.2	13.2
Atmospheric scientists, meteorologists	7.4	9.7
Statisticians	6.2	8.9
Civil engineers	3.8	8.2
Chemists	4.1	7.1
Earth scientists	3.3	5.0
Chemical engineers	7.0	5.0
Other social scientists	1.1	4.7
Medical scientists	NA	4.3
Economists	3.5	4.3
Biologists	1.5	3.8
Sociologists, anthropologists	NA	3.8
Biochemists	NA	3.4
Psychologists	0.4	3.1
Mining engineers	4.3	2.7
Petroleum engineers	3.4	0.9
Agricultural scientists	1.5	0.8
Urban and regional planners	0.8	NA

Regional Concentrations of Defense-Dependent S&Es: Overall, the pattern of defense spending has been heavily skewed toward the "gunbelt." Over the postwar period, the industrial heartland of the East, North Central, and Middle Atlantic have progressively lost out to New England and the Pacific regions, while most other regions have improved their positions modestly. Both in World War II and in the postwar period, substantial government-supported additions to industrial capacity, plus procurement contracts and R&D infusions, have helped to build agglomerations of defense-oriented activity in these latter regions. In addition, the evolution of commercially oriented segments of electronics, computing, and aircraft in the same locations has helped to solidify and diversify these gains.

Similarly, the SSE reveals that there has grown a corresponding concentration of scientists and engineers in two regions of the United States: Pacific Coastal states and in the Middle Atlantic. These centers of what has sometimes referred to as "bicoastal America" together had about 38% of all science and engineering jobs in the country in 1986. Within selected high-tech occupations, we see a clear difference between aeronautics (the West's strongest high-tech occupation) and electronics (the East's strongest). This reflects the heavy concentration of aerospace companies from Seattle to Los Angeles, whereas East Coast firms, especially in New England, focus more on electronics.

These regional variations are reflected in defense-dependent job shares within occupational groups. In other words, the likelihood that a scientist's or engineer's job is supported by the military will vary significantly depending on the region of the country. Aeronautical and astronautical engineers in the southern and western United States are more likely to be defense dependent than those in the east. Curiously, the Pacific region is relatively strong in the non-defense side of this occupation, perhaps due to the large commercial aviation operations at Boeing in Seattle and McDonnell-Douglas in Los Angeles. Among electrical engineers, we see that the South Atlantic and Pacific have a particularly high share of defense support. New England also shows a relatively high defense share in this occupation. More electrical engineering jobs in the traditional manufacturing belt of the Middle Atlantic and East North Central, stretching from New York through to Wisconsin, are supported by civilian markets.[70]

Defense versus Civilian S&E Salaries: Do defense-supported S&Es make higher wages than their civilian counterparts? For where we have the data, the answer is yes. Regardless of the measure of income used, defense-oriented scientists and engineers did better than their counterparts. The former group reported an average professional income of $46,011 for 1985, compared with $38,946 for those not supported by defense spending. The nondefense scientist or engineer made only 85% of what his or her counterpart on the defense side did.[71] This income differential grew dramatically during the Reagan defense buildup. All S&Es appear to have benefited, as their incomes on average jumped 44% in just four years. But defense-supported workers did much better. In 1981, the mean professional income of nondefense workers was 92% of the defense mean. In four buildup years, it fell to 85%. Apparently, as the demand for personnel from defense-oriented sectors grew, it pushed up wages and salaries to create a larger differential. Thus despite the fact that younger workers were disproportionately filling new defense-created job slots in the 1980s, tight demand pushed up incomes for those in the older cohorts who continued to work on defense projects.

Gaps exist within occupations, too. Aeronautical engineers not working on defense-related projects made only 93% of what their defense counterparts made in 1981, and this decreased to 84% in 1985. The latter enjoyed income growth of nearly 50% over the four-year period, while those not supported by DoD/NASA had increases of only a third. The gap was biggest for mathematicians. Those who did not work on defense projects made only three-quarters as much as those mathematicians who did. Widening gaps also existed within other defense-oriented occupations—physicists, astronomers, electrical engineers, and computer scientists. In most nondefense-dominated occupations—mechanical, mining, or industrial engineers, chemists, biologists or biochemists, and even psychologists—workers made more when working on defense projects than for other employers.

Military Dependency in Academia: As in industry, defense spending does not impact the academic R&D community uniformly. Overall, the Department of Defense supplies 12%-14% of all federal funds to R&D to universities and colleges and about 11% of all research obligations. In contrast to most other federal R&D obligations to academia, DoD's R&D funding is weighted more heavily toward applied research and development—which accounted for 22% and 32% of DoD's R&D support, respectively, in 1988—than toward basic research. The DoD provides to academic institutions only about 10% of federal funds for basic research, slightly more for applied research, but 43% of the development funds.[72]

DoD's pattern of R&D funding is also skewed toward a select set of disciplines (Table 9). We see that, in 1988, the DoD provided well over 40% of federal funds to universities for research in computer science, and metallurgy and materials, electrical, aeronautics and astronautics, and mechanical engineering. Mathematics and the environmental sciences are also major benefactors of DoD funds. If we look only at the research defense dependency of the hard sciences and engineering (*sans* life, social, and psychological sciences), the DoD's share of all federal funding to these disciplines jumps to 30%. If we include DOE and NASA support, whose missions overlap considerably with that of the DoD, these figures go up significantly, especially for the physical sciences.[73] Government R&D support also tends to be highly concentrated in a relatively small number of institutions. Of 3400 higher-education institutions in the U.S., the top 100 spent 83% of the total funds.[74] Pentagon agencies, in particular, tend to put most of their money into a few "centers of excellence." For example, of the nation's 200 doctorate-granting university departments in computer science, just *five* received nearly 80% of DARPA's academic funding for that field from 1976 to 1989.[75]

Tying DoD R&D funding to academia to the defense dependency of S&Es working in these institutions is somewhat tricky. Academics generally do not depend on external research funds for their livelihoods—base salaries are usually paid for by the employing institution—though their ability to raise such funds can affect various aspects of their professional life, including their ability to pay for various research expenses (including equipment) and graduate students, the size of their teaching loads, summer salaries, and their prospects for promotion and tenure. Plus, only about 30% of academic S&Es—28% of scientists and 34% of engineers—engage in R&D or the management of R&D as their primary work activity. More than half, however, consider teaching their main activity (Table 10). But, it is likely that many of these also do some research, just as most researchers probably do some teaching.

Federal/DoD funding more than likely affects doctoral S&Es to a greater extent than nondoctoral researchers and teachers in academia; some of the latter, though, may be employed in government-funded research projects under doctorate-holding principal investigators. Doctoral S&Es represent roughly 30% of all S&Es employed in academia and more than a third of all academic S&Es primarily engaged in research. About 35% of the academic doctorate-holding S&E population primarily engage in or manage R&D, while about half emphasize teaching; again, many, if not most of the teachers probably engage in some research, at least, to a larger extent than the nondoctorate teachers.[76]

The distribution of DoD R&D funds across academic disciplines probably affects S&E defense dependency in roughly the same proportion across the fields of academic doctorate-holding S&E population that does R&D and teaching. For example, from 50%-60% of the 38% of the doctorate-holding electrical engineers who engage in research, and to a lesser extent of the 53% who primarily teach—roughly, between 500 and 1000 researchers—may be considered defense-dependent (see Tables 9 and 10). The academically employed electrical engineers without doctorates or who do not participate in research or teaching are less affected by government or DoD funding.

Despite its relatively small size, however, the doctorate-holding research population in academia includes the brightest and most creative members of the overall S&E work force. Along with government R&D funds, they tend to be concentrated at the best and most important research universities in the nation, which are also responsible for turning out the next generations of top S&E talent employed in industry, government, and academia. Major cuts in R&D funding to universities would therefore have a detrimental effect on the nation's science and technology in much greater proportion than the numbers reflect.[77]

Military S&Es in the Federal Government: Although S&E growth in the federal government lagged behind that in industry and academia between 1976 and 1984, it rose at a faster rate than in these two sectors between 1984 and 1986, partly as a consequence of the defense buildup and partly because of relaxed hiring restraints.[78] Federal S&E employment was an estimated 412,000 in 1988, of which 48% were scientists and 52% were engineers.[79] Of this group, the DoD employs roughly half—including one-fourth of all federal scientists, two-thirds of engineers, and 60% of computer scientists. Another 31,000 or so S&Es—two-thirds are engineers, one-third are scientists—serve directly in the armed services.[80]

S&Es and the Prospects for Conversion

Vulnerability of Defense-Dependent S&Es

How vulnerable are S&Es to cuts in defense spending and programs? Jacques Gansler documents the "basic instability" of the defense labor force as a result of the broad fluctuations in the defense budget.[81] An aircraft plant in Ft.Worth, Texas, which is owned and operated General Dynamics and considered by Gansler to be a typical defense facility, has experienced wild swings in employment. Mary Kaldor writes that "between 1968 and 1976, defense-related employment fell from 3.2 million to 1.7 million; nearly half the total industry manpower lost their jobs."[82] The employment at the General Dynamics plant, fell from 30,000 to 7,000 in this period. S&Es are not shielded from these downswings in the defense procurement cycle. "Many engineers left the industry never to return," Kaldor adds. In California, a regional recession occurred with defense cutbacks in 1963-1964. Some 30,000 engineers were laid off, mostly in Silicon Valley and the Los Angeles basin, and had considerable difficulties finding re-employment.[83] In the Viet Nam builddown, unemployment among engineers increased from 0.7% in 1967 to 2.9% in 1971, an increase much greater than that for all other professions.[84]

The difference between past cycles and the current situation, however, is that for some industries and firms there may not be another upswing—something they once were able to count on. What may make the adjustment particularly difficult in the 1990s is the likely large-scale nature of engineering layoffs. The Institute of Electrical and Electronics Engineers predicts that 55,000 defense-industry engineers could lose their jobs by 1995, on top of thousands laid off already. Job insecurity could become a more common experience for these workers, as they bounce around in a more competitive labor market. Donald Hicks, a defense-industry consultant and former Pentagon and Northrop official, put it succinctly: "An awful lot of engineers and scientists will be driving cabs."[85]

Since military spending is heavily concentrated in certain industries, occupations, regions, and cities, peace threatens considerable displacement and community distress unless a concerted effort to help plan and smooth the transition takes place. With cuts looming, as many as 1 million civilian workers could lose their defense-related jobs between 1989 and 1995, including 830,000 in the private sector. These losses would be heavily concentrated by industry. In communications equipment, employment is predicted to decline from 349,600 in 1988 to 213,400 in 1994; in guided missiles, from 134,900 to 105,000, and in shipbuilding, from 102,300 to 61,700. Add to that the 320,000 to 550,000 active-duty military and DoD personnel who will be shed, and the country faces structural unemployment comparable to that which occurred in the early 1980s in the industrial heartland.[86]

Which industries, firms, occupations, and communities will be hardest hit will depend on how rapidly the defense budget declines and how the Pentagon decides to restructure the nation's military forces in the post-Cold War, post-Gulf War period. The war against Iraq confirmed for government planners the value of certain types of military assets. But it will not prevent the steady erosion of the defense budget or the restructuring process over the coming decade. DoD's projected plans through 1997 would substantially reduce the number of service personnel and military installations, eliminate or cut back on a number of major strategic weapons programs—including a downsizing of the Strategic Defense Initiative program,[87] modernize only selected conventional systems and technologies, and reduce spending for technology-based R&D.[88]

The impact of this plan on the S&E labor force will depend on how it is actually implemented. Workers employed in firms involved in making conventional high-tech missiles and certain classes of aircraft may be more secure than those who build large strategic nuclear weapon platforms like the Trident submarine or M-X missile. It will also depend on whether the Pentagon will let the shakeout in aerospace and other defense industries take place without interference or if it will try to keep its main contractors in business through its practice of rotating winners and losers—which it seems to have recently done with the large F-22 fighter aircraft contract to a Lockheed-led team. Uncertainty about winners and losers in the military-industrial system is enhanced by the pork-barrel tendencies of Congress, which often votes back into the budget programs originally cut by the DoD. Within this environment, a more precise evaluation of the vulnerability of defense-dependent S&Es and their prospects for conversion will have to take into consideration two main factors: labor market supply-demand dynamics and structural and "cultural" differences between civilian and military production activities.

Table 9

DoD Support for University Research by Discipline[89]

FIELD	DOD SUPPORT FOR UNIV RESEARCH [MILLIONS OF $ 1990]		% SHARE OF FED $ FOR UNIV RESEARCH-1988	
	1978	1988	DOD	NSF
Total All Fields	**445**	**785**	**11**	**17**
Life Sciences &Psychology	56	99	2	5
Physical Sciences	80	102	11	38
Astronomy	5	5	6	37
Chemistry	23	45	14	38
Physics	48	52	11	33
Environmental Sciences	94	99	22	54
Atmospheric sciences	14	27	22	47
Geological sciences	12	29	25	48
Oceanography	62	42	25	69
Mathematics	22	36	28	52
Computer Science	39	124	61	26
Engineering	152	321	44	27
Aeronautics & astronautics	29	46	54	—
Chemical	6	3	5	60
Civil	2	11	26	70
Electrical	57	86	57	37
Mechanical	25	30	37	39
Metallurgy & materials	32	126	67	13

Table 10

Employed Academic S&Es and Academic S&Es with Ph.D.s—Selected Fields

FIELD	ACADEMIC S&ES-1988[90]			ACADEMIC PH.D.S-1987[91]		
	TOTAL ACAD (000S)	% TOTAL R&D**	% TEACHING	TOTAL ACAD (000S)	% TOTAL R&D**	% TEACHING
Total all fields	**742**	**29**	**51**	**219**	**35**	**49**
Total Scientists	**624**	**28**	**52**	**195**	**35**	**49**
Physical scientists	78	32	61	30	42	50
Mathematical scientists	75	12	78	14	22	68
Computer specialists	48	20	38	6	29	48
Environmental scientists	18	46	46	8	47	44
Oceanographers	2	95	5	1	62	23
Atmospheric scientists	2	76	14	1	73	18
Life scientists	164	51	37	66	53	7
Total Engineers	**118**	**34**	**44**	**24**	**38**	**49**
Aeronautical/ astronautical	4	38	62	1	56	44
Electrical/ electronics	30	39	43	4	38	53
Materials	5	58	38	2	53	37
Mechanical	24	33	43	4	29	60
Other engineers	28	33	42	6	45	37

S&E Supply-Demand Dynamics

The potential ability for defense S&Es to make the shift from military to civilian jobs depends on the market demand for his/her field of specialty, in general, and his/her skills and experience, in particular. To facilitate the transition of large numbers of S&Es, the cuts in defense demand will need to be matched by comparable off-setting growth in the civilian sector. The demand projection will also have to be matched to supply factors. The numbers of new S&E graduates, attrition and retirement rates, and the age, occupational composition, and location of the pool of S&E workers in the labor market are factors that will affect the ease of transition for workers laid off from defense-related jobs.

S&E Demand Projection: The projected demand for S&Es through the year 2000 looks very positive. Economic growth, changing technology, and competitive challenges and the need to replace workers who leave the S&E work force because of death, retirement, transfer to other occupations, and emigration, all contribute to the need for new workers. According to a National Science Foundation study, between 1988 and 2000, almost 600,000 additional jobs for S&Es are expected to be created in private industry, which is four times the projected growth rate for the total industrial labor force. [92] Among the factors driving the increase in S&E demand in industry are the general growing need for workers with more education, and the need to remain competitive, which "will require an increasing number of S&E workers to update product designs, explore more cost-effective technology of producing goods, and develop new products."[93]

Three-fifths of this increase is expected to be in the services-producing sector. The S&E growth in this sector is part of the anticipated overall employment increase in the sector that the NSF says will be about 13% higher in the year 2000 than in 1988.[94] Business-related services is projected to be the fastest growing industry within the service sector, with an expected rise of over 65% in S&E positions—jumping to 64% of all service sector S&E jobs—by the year 2000. The demand for the services of independent research laboratories and of management and other consulting firms, in particular, will be very strong.

Similar to the trend observed between 1980 and 1988, there will be little, if any, growth in the goods producing sector's total employment. Nevertheless, growth in the output of most manufacturing industries is expected to continue at near current rates despite the lack of concurrent growth in employment. Meanwhile, the general occupational shift away from production and assembly-line toward professional, managerial, and technical occupations will continue. Hence, while employment in non-S&E jobs are expected to decline slightly over the 12-year period, S&E jobs are expected to rise by 256,000 or 32%.

The demand for certain occupations, especially computer scientists and electrical/electronics engineers, will grow at especially high rates over the same period. According to a 1989 survey of industrial establishments by the NSF, "nearly 46% of firms in major S&E employing industries reported shortages in one or more S&E fields." Those fields with the highest reported levels of shortages were: electronics engineers, at 22% of employers reporting; chemistry, at 21%; electrical engineering, computer engineering, and computer science, each at 20%; and chemical engineering, at 18%.[95] The demand for computer specialists is expected to grow at the fastest rate and in the largest numbers, half in business services and data-processing services. The remainder will be spread throughout industry as computers are used more extensively by industrial concerns. New business and defense computer applications will continue to be primary sources of requirements for computer specialists. Electrical/electronics engineers are projected to have the next biggest absolute and relative gain, roughly divided equally between the goods and services-producing sectors.   Demand will be the greatest in the manufacturing industries producing communications equipment, computers, and other electronics equipment, and in the engineering and architectural services firms servicing these industries.

Academia will also continue to have high demand for qualified S&Es. Engineering schools in the U.S. report serious and persistent shortages of faculty. In 1986, almost 9% of authorized full-time engineering faculty positions went unfilled. The shortage of engineering faculty is linked to both a long-term decline in engineering doctorates to U.S. citizens and the increasing demand for doctorate-holders in the industrial sector.

There is one indicator of S&E labor market conditions, however, that is not so positive: the high-technology recruitment index (HTRI), which measures the amount of advertising space devoted to recruiting scientists and engineers.   In the last five years, the HTRI has gradually declined, registering the lowest level of activity since 1983. Over the short term, then, the demand for S&Es is relatively low, which could have an equal or greater bearing on the future job possibilities for soon-to-be unemployed scientists and engineers than the long-term trends.[96]

S&E Supply Projections: The S&E supply is contingent upon the graduation for new S&Es from the nation's universities and colleges. Other direct flows into the work force include reenters, immigration, transfers, and separations from the S&E work force. All these flows, the NSF study notes, "are highly sensitive to S&E supply/demand conditions, including such factors as economic growth; defense and R&D spending; the demographics, interest and capabilities for the population which affect the number of new entrants to and attrition from stock; legislation and policy affecting the entry of foreign students and immigrants; and policies and laws regulating retirement ages; etc."[97]

S&E degree production has grown steadily over the last decade. S&E baccalaureate awards grew at about 2% a year, reaching 324,000 in 1986. The field distribution of baccalaureates has shifted, significantly favoring the engineering and mathematical sciences (primarily the computer sciences), while degrees in the life and social sciences have declined the most. After declining in the late 1970s S&E masters degree production rose sharply in 1982, rising to an all-time high in 1986 of more than 62,500 S&E masters awards granted. Ph.D. awards have also grown steadily since 1978. In 1988, roughly 20,300 doctorates were awarded representing 61% of all doctorates. The growth in engineering and physical sciences made the largest contributions to S&E doctoral degree growth since 1982. Mathematical and computer sciences doctorates also grew at a rapid rate. But, the NSF reports, "all the growth in the number of S&E doctorates awarded since 1978 is accounted for by rapid increases in the number of foreign students receiving such degrees." Foreign students received 49% of all engineering doctorates in 1988 and 27% in the sciences.

Despite these growths, there is concern that the flow of entrants into the S&E labor force may slow down in the future. First, the size of the 18-24 year old population, from which new S&Es would be recruited, is expected to decline in the early 1990s. Second, retirements are expected to increase as larger numbers of experienced S&E workers reach 55 or older.[98] In addition, the changing ethnic and gender mix of college-age population, which in the future will have higher percentages of women and minorities—groups that traditionally have had low levels of participation in S&E—also creates concern about future S&E supply.

Prospects for Defense S&Es

Based on these projections, the prospects of defense S&Es for finding new jobs, upon reentering the labor market as a result of military cutbacks, look promising. Demand for new S&Es in the industrial work force is projected to grow significantly over the next decade, while supply, though it will still grow, is expected to slow. Hence, S&E shortages will continue both in industry and academia. Employers have therefore become more open to hiring experienced S&Es in greater numbers. The NSF reports, for example, that "across all fields, 80% of the employers of scientists and 75% of the employers of engineers reported plans to hire experienced personnel rather than new college graduates."[99] In principal, at least, former military S&Es may find themselves in demand in a labor market characterized by serious shortages.

Some caveats should be placed on these demand projections, however. First, the NSF notes that the expected growth in S&E demand, while significant, may not match past increases. The reason for this, according to the NSF, is "the overall slowdown of the labor force, total employment and GNP growth expected in the 1990's."[100] Also, the downturn in defense spending, especially in R&D, is not anticipated in the NSF's original scenario, upon which it based its projections. NSF's rerun of the projections based on proposed additional cuts showed that S&E demand would "increase at a slightly lower rate than" earlier predicted.[101]

Also worrisome is a significant slowing down in overall R&D investment in the nation by both the federal government and private industry in recent years, which some fear could hurt U.S. competitiveness.[102] This also could cut into future S&E demand growth, especially for doctoral and other R&D-oriented S&E personnel. In addition, the main projected growth is expected to be within the service sector, not in the manufacturing areas, where the defense cutbacks will be the most immediately and acutely felt. The decline in defense would probably slow demand in the goods producing industries to a far greater extent than in the services sector. Hence, there could be a very serious mismatch between the industrial orientation and experiences of the S&E workers who will be thrown into the job market by military cuts and the requirements of the jobs in the civilian sector that will be most available—a point to which we will return later.

Joshua Lerner's findings on S&E mobility, reported in his analysis of the Survey of Scientists and Engineers, are suggestive here. He detected a fairly high degree of shifting between defense and nondefense work in the 1980s. The movement appears to have gone in both directions, despite the considerable defense buildup. Of those S&Es who worked on defense projects in 1982, almost one in four sampled had shifted out of defense work by 1986, a surprisingly high share for a period of intense military buildup. However, the majority of them appeared to have shifted projects within firms rather than from one firm to another.[103] Rates of mobility were highest among those concentrated in industries with relatively lower degrees of defense dependency, computing and electronics in particular. Low mobility rates, on the order of 15 to 20%, were recorded for those in aircraft and ordnance. Younger workers had relatively higher rates of mobility from one sector to the other, while Lerner found those who stayed with defense work to be relatively older and "plateaued."[104]

S&E Concentrations and Supply-Demand Flows: If the defense S&E work force was more evenly dispersed throughout the overall S&E population than the supply-demand projections suggest, that readjustment of dislocated defense S&Es could easily be handled by ordinary market forces. But our projections of how easily the labor market can handle major layoffs of defense S&Es need to be tempered by Lerner's finding, the caveats we raised earlier, and consideration of a key characteristic of the defense-dependent S&E work force noted earlier: its unusually high levels of concentration within certain industries, firms, occupations, and geographical locations. These factors make modeling the supply-demand flows of S&Es somewhat "sticky." It is difficult to incorporate their effects into models that assume smooth, aggregate movements in the supply-demand functions, rather than the discrete step-functions and localized concentrations that actually characterize the dynamics of defense S&E employment. When layoffs do occur, due to large weapons cuts (or installation closures), the major contractors generally do not reduce their work forces by attrition; rather, as the wild employment swings noted by Gansler illustrate, they lay off large groups of workers *en masse*, over a relatively short period of time.

Hence, large blocs of S&Es of similar fields and experience, from the same industry and firms are thrown into same local labor pool, having to compete with each other to find jobs in a community that may or may not have growing civilian demand for their skills. This is compounded in periods of major defense downturns affecting several firms within defense-dependent communities. As long as the military budget is secure, workers could move over to other firms where new contracts were won. But when there is less likelihood of new follow-ons, and several local firms employing the same types of workers are also cutting back on their personnel, then regional concentrations make it more difficult to re-employ workers in the same area. These multiple layoffs therefore put additional pressures on local community resources to provide adjustment assistance.

The health and diversity of the industrial bases of locales where the defense cuts are made will be an important factor that determines how difficult it will be for defense S&Es to find other work. But unless the civilian sectors of these communities are expanding at unusually robust rates, which has not been the case in the U.S. in recent years, the mismatch in defense and civilian industry types and occupational composition will make readjustment of the affected S&E work force very difficult. First, as noted already, the defense industry is part of the durable goods sector, which is not expected to see a rapid increase in S&E demand in the near future. Thus, workers from several manufacturing-oriented occupations, such as aeronautical and astronautical engineering, physicists, mechanical engineering, and certain specialties in electronics/electrical engineering, for example, may find it especially hard to make the transition to jobs in the services-producing sector. Certain occupations, such as computer specialists and other specialties in electrical/electronics engineering, which will be in high demand in the service sector, may suddenly be in temporary oversupply, until S&Es from these fields are able to make their way to communities or other local businesses where the demand for their skills is high.

The second problem facing laid-off defense S&Es is that even if a strong civilian demand exists, civilian industries will not need their services in the numbers that were formerly employed in defense plants. Seymour Melman argues that in the aerospace industry, for example, "there are seventy-eight administrative, technical and clerical employees for every hundred blue-collar production and service workers." "Civilian enterprises need engineers and technicians," he adds, "but none of them require these people in the proportions found at [a typical aerospace firm] where more than a third of the employees are engineers and supporting technical workers." Hence, he warns that only "about one-third of these engineers, if retrained for civilian engineering would be employable in a converted enterprise of a similar size."[105] It will be equally difficult for them to find jobs in the less S&E-intensive civilian sector in their local communities, where they are forced to compete with their former colleagues, as well as new graduates entering the field. Many, as Melman notes, "would have to either change occupation or change job location or both."

Skills Mismatch and "Cultural" Divergence

Aside from the overall supply-demand dynamics, it remains a crucial question whether the skills and experience of those leaving the defense sector will sufficiently match those required in order to successfully compete in the civilian marketplace. The large differences between defense and civilian-oriented firms, or even between the defense and civilian divisions within the same company, in work organization and environment, and their divergent technological design cultures, have been documented by a number of writers in the past (especially see the works of Melman, Dumas, and Ullmann). It is important not to overgeneralize here. These differences are probably greatest in the aerospace, ordnance, and very specialized military electronics firms, but much less pronounced or nonexistent in the general-purpose components industries that supply defense firms, for example.

Defense firms are known for their excessive levels of bureaucracy—which evolved largely in order to deal with their equally bureaucratic customer, the Pentagon—which is top heavy with administrative, technical, and clerical personnel. A second difference is in the type of product made. Military production tends to emphasize products that are large-scale, customized, use highly specialized and sophisticated technologies, and are produced in limited quantities. This contrasts with the standardized, mass-production orientation of most civilian industry, even at the high-tech end. As a consequence, civilian design requirements stress standardized, high-quality goods produced in quantity and low cost, as opposed to military design requirements, which stress very high-performance and customized specifications (i.e., milspec), with cost criteria only a secondary concern. Military engineering design also operates within a longer time-to-production cycle than civilian engineering, where "labor learning"—which comes out of many iterative design-production-marketing cycles over time—is subsequently limited.

A third major difference between the civilian and military sectors is the security apparatus that the latter is forced to overlay on its design and production operations. The security administration within defense plants not only adds an extra layer of bureaucracy, but, by limiting and channeling workers' movements and communication, it contributes to the fragmentation and overspecialization in the work process; defense workers must abide by a "need to know" rule, which makes it difficult to talk among themselves about their work, much less share the day's work with their family and friends.   Information flows within defense firms are therefore even more inhibited than in commercial companies, which, among other things, helps to stifle employee innovation.   Because of security restrictions, S&Es from defense companies are not as able as those in civilian industry to capitalize on ideas and innovations developed in the course of their work to initiate new start-up businesses.

Barriers to Transferability: As a result of the structural and cultural differences between civilian and military enterprises, the job mobility of defense S&Es and the transferability of their skills to civilian jobs are greatly circumscribed.   In a study of the transferability and conversion of defense occupations by SRI International for the Arms Control and Disarmament Agency in 1967—one of the few such studies ever to be done—several *barriers* to transferability were identified.   These include the following:

- "Defense engineers are extremely specialized, to a degree not found in commercial areas.   Commercial engineers are product-oriented, while defense engineers tend to specialize in more narrow technical fields and do not usually follow a product through from start to finish.
- Defense engineers design for maximum performance, regardless of cost, while cost is a major factor in commercial design.
- There are more electronic and aeronautical engineers in defense industry by far than in commercial industry.   In general, there are more engineers per sales dollar in defense than in civilian work.   This creates a structural problem of lack of available jobs in the commercial sector.
- Most defense engineers are more highly paid than their commercial engineering counterparts and are thus reluctant to transfer to lower-salaried commercial jobs.
- The defense industry makes more extensive use of engineers in systems analysis and system design.
- Certain defense engineers, such as aeronautical and documentation engineers, do not perform conventional engineering functions but are employed to meet military requirements for reports and specifications. They have major problems transferring because of the lack of comparable jobs in civilian industry."[106]

Mobility Obstacles: The mobility obstacles for engineers—such as overspecialization, and one not yet mentioned, technical obsolescence—are reflected in other anecdotal and formal reports as well.[107]   Defense managers report considerable difficulty in shifting engineering and science personnel from military-oriented to commercial projects within the company. Those working on defense become narrowly specialized in a particular technology and find it difficult to think in the more pragmatic and wide-ranging way required on the commercial side.[108]   Some companies, like Boeing, have permanent programs to encourage lateral moves from military to commercial divisions and back. But many complain that adjustment for older, more narrowly specialized engineers is a particularly big problem.

Mobility outside of the firm, especially to the civilian high-tech sector, is even more difficult. Mobility is fairly high for the youngest engineers, some of whom have deliberate strategies of joining up with contractors who will pay to complete their education before they move on. Older workers have a harder time making the switch. Marion Anderson quotes Bill Bradford, Executive Secretary of the Seattle Professional Engineering Employees Association, as saying[109]:

> I think that the major problem is keeping up with the state of the art, whether it's aero-astronautic, or computers, or industrial or whatever. For instance, if a man has worked for fifteen years on radar tracking, and a new technology comes along, his skills are obsolescent. He's fifty. This is a bad problem.

Middle-range tech people, according to one engineer, even those who "got a little juice left to them," have a hard time moving to civilian firms. Even in Silicon Valley, where commercial job openings abound, engineers who have worked for large defense contractors like nearby Lockheed Missiles and Space, the single largest Silicon Valley employer with a staff that has fluctuated between 20,000 and 30,000, find considerable prejudice against them in local labor markets. Some prospective employers worry that engineers from a big military contractor will quit their new jobs to rejoin the original company once contracts flow again. More commercially competitive employers worry that laid-off defense engineers and scientists will be overspecialized and untutored in cost consciousness.[110] Such firms are looking for young hot shots who are more malleable, who can get easily plugged into the commercial culture. As a result, discrimination against the restless military aerospace engineer is strong.[111]

The glut of engineers on a depressed labor market therefore could drive many to accept inferior jobs or try new occupations. For some, it has happened already. Forty-two-year-old computer systems engineer Bruce Hill was laid off in February of 1988 at Lockheed's Austin, Texas, subsidiary. He subsequently worked at Raytheon in Boston, E-Systems back in Texas, and then FMC Corporation in Silicon Valley, where he works on the Bradley armored vehicle. Hill's income has plunged by one-third, and he lost tens of thousands of dollars on his home when forced to sell it. Hill, according to the reporter who interviewed him in June of 1990, has decided to try working for the government as a safety engineer if he's laid off again. It doesn't pay well, but it's sheltered from layoffs.[112] But even jobs with the government may be hard to come by. In early 1990, Defense Secretary Dick Cheney announced a civilian hiring freeze, noting that the Pentagon would probably eliminate more than 50,000 jobs over the next year.[113]

Engineers in Conversion Planning

Most of the prior discussion focused on the conditions affecting the prospects for occupational conversion of defense-dependent engineers and scientists. Another option for S&Es is to join in attempts, with blue and pink collar workers, plant managers, community leaders, and elected officials, to keep their plants open by converting from military to civilian production. Engineers and scientists have played important roles in the limited number of conversion efforts that have been attempted so far, and can and should play key roles in any such efforts in the future. It is their job, after all, to do the research, development, and design for the products that their firms produce and sell. They have the technical expertise and experience, which, when coupled with the tacit and experiential knowledge of shop-floor workers and the administrative, organizational, financial, and marketing skills of managers and outside consultants, constitute the basic ingredients for producing a conversion plan.

Engineers were at the forefront in the development and promotion of the famous Lucas Aerospace Corporate Plan in Great Britain during the 1970s. Lucas Aerospace, which made aircraft parts for the military and several other products, was undergoing a major reorganization, instigated by its parent conglomerate, that threatened to cause major layoffs in its nearly dozen plants spread throughout the English countryside. Leaders from the main unions representing engineers and designers at these facilities subsequently met with shop-floor union leaders to form the Lucas Aerospace Shop Stewards Combine. The Combine then proceeded, with a little outside help from university scientists, to devise a corporate plan, which identified about 150 products the company could pursue rather than shut down its facilities. Although this effort only met with limited success—some jobs were saved and "redundancies" (i.e., layoffs) slowed down—it became a landmark effort that inspired and informed many of the bottom-up conversion projects undertaken by local groups, labor unions, and conversion organizations during the early 1980s in both the U.S. and Western Europe.[114]

The existence of unions representing S&Es facilitated the involvement of this workforce in the Lucas effort. In the U.S. very few engineers and scientists are unionized, however. Interestingly, many of the handful of such unions in this country are located at major military aerospace plants—at Boeing, in Seattle, and McDonnell Douglas, in Long Beach, CA, for example. The engineering union at the Long Beach plant participated in an effort during the early 1980s spearheaded by the United Auto Workers Local 148—assisted by the Center for Economic Conversion and the State of California's Department of Commerce—to explore "new products for idle plant utilization," such as light rail mass transit vehicles, cogeneration equipment, and commuter aircraft. At that time, the Douglas plant was operating only at 30% capacity and over 10,000 of its workers—

blue and white collar—were unemployed. Its fortunes reversed themselves only when Douglas got a multibillion dollar R&D contract for the C-17 Air Force transport.[115]

Corporate Examples: In the late 1960s and early 1970s there were a number of notable corporation-initiated "conversion"—in reality, diversification—efforts that met with mixed success. In each case, varying degrees of in-house engineering talent were employed in the planning and design processes, and some found new employment in new divisions producing civilian products. Kaman Corporation successfully harnessed some of the technical talent and knowledge of workers in its military helicopter division to design a successful line of guitars (Ovation). CEO Charles Kaman had recognized that some of the principles for reducing and controlling vibration in helicopter blades could be used in designing a superior type of guitar. Kaman also employed some of its existing knowledge base developed through the construction of helicopters in a second successful diversification, this time into bearings.[116]

Acurex Corporation, originally an aerospace firm, also was able to diversify into environmental services and solar energy products, employing the technological expertise it had developed in doing work for the defense community. In this case, about 20 to 30 engineers from the firm's aerospace group helped to start the environmental business, although crossover declined as the environmental and energy businesses grew. Robert DeGrasse notes in a report for the Department of Defense, that "those personnel transfers which did occur typically involved young engineers seeking better opportunities." Nevertheless, he credits the Acurex experience as representing "an important example of a Defense contractor redirecting its resources for non-Defense purpose."[117]

The Boeing-Vertol and Rohr efforts to diversify into mass transit vehicles are also well-known cases. Both firms ran into unforeseen problems, which eventually led to their retreat from the mass transit market. Each drew upon engineering talent from their military divisions—Boeing-Vertol was making Chinook CH-47 helicopters for the Army and Rohr was a major aerospace contractor—to design and produce mass transit vehicles for large urban clients. About 75%-85% percent of the engineers and supervisors who worked on Boeing's transit projects had been shifted from the company's defense production. Rohr applied aerospace skills, such as systems engineering, structural design, quality control, and fabrication, to its new civilian ventures.

One of the problems both companies had was underestimating the technical challenge involved in producing new railcar technologies. Boeing had essentially agreed to produce a new generation of trolley car. DeGrasse reports that "Boeing officials confirm that the company failed to fully anticipate the engineering challenges involved in the project. They also admit that, as producers of sophisticated military hardware, Boeing engineers and managers looked upon building transit cars as a technically less demanding task."[118] Similarly, a Rohr official conceded that Rohr had made a misjudgment in "offering to do things that were beyond the state of the art."[119]

Although the technical difficulties involved in shifting from aerospace to civilian production were key factors in the failed conversion efforts of these companies, additional determinants included the lack of a strong mass transit market, inadequate government policies, and other market factors. If both the market had been stronger and government leadership more effective, Boeing, at least, may have been able to remain in the mass transit business. DeGrasse observes, for example, that after Boeing's failure in providing reliable railcars for Boston's transit system, "Boeing's experience in providing cars to the Chicago Transit Authority provides... evidence that the company evolved into a competent supplier of mass transit vehicles."[120] Its cars in San Francisco also performed well.

But, even if it had succeeded, only a small portion of Boeing's S&E work force would have been affected. While production of the Chinook helicopter had reached a peak of 13,000 employees during the Vietnam War, the new transit division never employed over 550 personnel, including 150 engineers and 400 production workers. Only small numbers of defense engineers were able to find work in the new civilian enterprises of the other companies as well. These experiences are compatible with one of the most important conclusions of the SRI report cited earlier, that "individuals are transferable, particularly if they can be brought gradually into a functioning commercial unit so that they can learn on the job by absorption. Organizations are not transferable."[121] Individual S&Es can make the transition—if certain difficulties are recognized and steps taken to overcome them. But, the problem is substantially compounded for converting large blocs of S&Es, which is the reality that will have to be faced when major cutbacks occur.

Policy Issues—The Need for New National Initiatives

It has sometimes been argued, "Why should the government help alleviate the adjustment problems of defense workers—particularly S&Es—which is really an elite group comprised primarily of well-paid, white males, when other dislocated workers in the civilian sector get hardly any help at all?" There are a few ways to respond to this point. First, the hardship of one group of people is not a good reason not to help another, even if more privileged, when they are in trouble. In reality, any conversion or adjustment assistance for S&Es should be extended to *all* eligible workers in the economy. Second, the plight of the defense worker was created by government policies, not by normal market forces. The modern S&E occupations and work force were in large part the creation of the Cold War, and the potential large-scale dislocation that defense workers now face is the product of the winding down of that same "war." The government therefore has the responsibility for aiding the transition for defense-dependent workers, firms, and communities so that they can rejoin the civilian market.

Finally, a more positive reason to support S&E conversion is to preserve the enormous public investment that our society has made in educating, training, and employing the S&E work force. The S&E work force should be considered a national asset and resource, which we will need to employ to solve many of our most pressing economic, social, and environmental problems in the coming decades. The real problem therefore is not how to help engineers and scientists readjust to civilian work, but how to re-deploy and redirect our scarce scientific and technological resources to meet the needs of our society and of the world.

We saw that making the transition for defense S&Es will not be an easy matter. With just some extra assistance *individual* S&Es can make the adjustment without that much difficulty. But given the large concentration of defense S&Es in industries, firms, occupations, and communities, on one hand, and the numerous structural and cultural "barriers" to transferability on the other, ordinary market forces will not be sufficient to handle the problem. Some form of government assistance will be essential. The various retraining and adjustment assistance legislation proposed by different people will be helpful, but are only one component of the overall policies that will be needed. Facility conversion legislation will also be important, but not sufficient. Facility conversion may be a viable option in some instances of major cutbacks at defense facilities, but not in all cases, and not all workers will necessarily benefit. In order to make both the individual conversion and industrial conversion more viable, some broader, more far-reaching policy initiatives will have to be instituted, which will stimulate effective demand for new investments in socially important sectors of the economy and create new jobs for *many* types of workers.

Such initiatives would include strong R&D programs designed to tackle critical scientific and technological problems and develop new socially and environmentally valuable products and processes—new pollution prevention technologies designed into advanced computer-aided manufacturing systems, for example. Readjustment assistance and conversion planning policies would be essential elements in these initiatives; they will be needed in order to ease the short-term adjustment problems of workers, businesses, and communities as we shift our resources to new civilian ends. Together, these policies and initiatives constitute what we call an economic development strategy for the nation, which we believe should be based on three principal national goals: economic stability, health, and environment. Aspects of such a program are discussed in more detail elsewhere.[122] Not only will this kind of policy—which, in effect, is a national conversion strategy—help direct our science and technology resources toward meeting important national objectives; it will also stimulate many new areas of R&D and science and technology innovation, creating the job opportunities that will be needed for the transition of scientists and engineers and newly graduated S&Es.

Notes

1 John H. Cushman, Jr., "Grumman F-14 Pact Is Canceled," *New York Times* (27 February 1991): D1.

2 Steven Pearlstein, "Other Bidders Settle for Smaller Bites of the Aerospace Pie," *The Washington Post National Weekly Edition* (29 April - 5 May 1991): 11. Quote is attributed to Lockheed chairman Daniel Tellep.

3 Eric Schine, "Defenseless Against Cutbacks," *Business Week* (14 January 1991): 69. Quote is attributed to David J. Wheaton, vice-president for planning at General Dynamics Corp.

4 The "gunbelt" is a half-moon-shaped defense perimeter, thickest in Southern California and north through Silicon Valley and Seattle and eastward through the Intermountain and Plain states, through Texas, Florida, Long Island, and New England, and conspicuously detouring the traditional industrial heartland. The documentation and existence and historical evolution of the "gunbelt" is presented in Ann Markusen, Peter Hall, Sabina Deitrick, and Scott Campbell, *The Rise of the Gunbelt* (New York: Oxford University Press, forthcoming 1991).

5 Jacques S. Gansler, *The Defense Industry,* (Cambridge, MA: The MIT Press, 1980), pp. 50ff. Gansler uses the term labor learning in "the traditional production-theory sense, to mean the cost reduction realized as a worker performs the same job repetitively, becomes more skilled at the task, and is therefore capable of doing the task in less time. This is usually calculated as a percentage reduction in unit cost as a function of the quantity of items produced." p. 291, n. 39.

6 Ibid., 50.

7 Ibid., 50.

8 Unless otherwise indicated, this figure, and those used in the text immediately following, are based on data reported in National Science Foundation, *U.S. Scientists and Engineers: 1988, Estimates,* Detailed Statistical Tables, NSF 88-322 (Washington, DC, 1988); Another NSF data source is National Science Foundation, *Science and Engineering Personnel: A National Overview*, Special Report, NSF 90-310 (Washington, DC, 1990). The latter data source contains S&E employment estimates for 1988 that are a little different from those appearing in the first source. For example, total S&E employment is reported here as only 5.3 million, scientist employment as equal to 2.6 million and engineers as 2.7 million. The proportions of S&Es with respect to different characteristics (e.g., primary work activity) will therefore be slightly different, depending on which data set is used. The NSF employs a series of surveys and modeling techniques to generate its figures. Discrepancies between the data sources probably reflect differences in the survey data NSF employed in the statistical model used to generate the different estimates. The reader is therefore cautioned that the 1988 year

data used here, in particular, should be considered only approximate or ball-park figures.

[9] An individual's primary work activity is defined as that activity on which one spends the largest proportion of time, but not necessarily full time.

[10] National Science Foundation, *Science and Engineering Personnel: A National Overview*, Special Report.

[11] National Science Foundation, *U.S. Scientists and Engineers: 1988, Estimates,* Detailed Statistical Tables, NSF 88-322 (Washington, DC, 1988): 1988 figures are estimates.

[12] NSF, *U.S. Scientists and Engineers: 1988, Estimates,* Detailed Statistical Tables, NSF 88-322 (Washington, DC, 1988): 1988e figures are estimates.

[13] Ibid.

[14] NSF, *Science and Technology Data Book, NSF 88-332* (Washington, DC 1989).

[15] NSF, *Science and Engineering Personnel: A National Overview, Special Report, NSF 90-310* (Washington, DC, 1990).

[16] National Science Foundation, *Science and Engineering Personnel: A National Overview, Special Report, NSF 90-310* (Washington, DC, 1990): Table B-18, pp. 88ff.

[17] Figures in thousands.

[18] Less than 1/2 of 1%.

[19] National Science Foundation, *Science and Engineering Personnel*, 19-20. While this percentage is up from 8% in 1976, this report notes that only about 1% of all employed women, as a proportion of the total work force, were working in S&E jobs in 1986, compared to 6% of all employed men.

[20] National Science Foundation, *Science and Engineering Personnel*, 20-24. NSF data taken from U.S. Department of Labor Statistics, *Employment and Earnings,* 34, No. 1 (Washington, DC: Supt. of Documents, U.S. Government Printing Office, January 1987): 179; Latinos (i.e., people with Hispanic origins; not shown in Table 3) are also underrepresented. While constituting roughly 6.6% of all employed persons and 3.3% of all those employed in professional and related occupations, Latinos represented only 2.1% of the S&Es employed in 1986.

[21] For example, an NSF report observes that "women scientists and engineers are more likely than men to work in nonprofit organizations, academic institutions, and state and local governments. They are less likely to be employed in industry or the federal government." National Science Foundation, *Science and Engineering Personnel,* 19-20.

[22] National Science Foundation, *Characteristics of Doctoral Scientists and Engineers in the United States: 1987,* Detailed Statistical Tables, NSF

88-331, (Washington, DC, 1988). Overall, in 1987, 15.9% of all employed S&Es, 19.5% of scientists, and 2.5% of engineers with doctorates were women. Blacks accounted for 1.5% of the total S&E work force with doctorates, 1.6% of the scientists and 1% of the engineers. Hispanics represented 1.7% of S&E doctorates in 1987.

23 Ibid.

24 Ibid. Another 4% work in nonprofits and 7% for the federal government.

25 Ibid.

26Harry Braverman, *Labor and Monopoly Capital* (New York: Monthly Review Press, 1974): 242.

27 Daniel Bell, *The Coming of Post-Industrial Society* (New York: Basic Books, 1976): 134-216. Bell reports that between 1930 and 1964 the overall work force grew by only 50%.

28 Professional workers, technicians, and related workers together may account for the largest bloc of employees in the United States by the year 2000. See Ronald E. Kutscher, "Overview and Implications of the Projects to 2000, *Monthly Labor Review* 10, No. 9 (September 1987): 3-9.

29National Science Foundation, *Characteristics of Doctoral Scientists and Engineers in the United States,* Detailed Statistical Tables, NSF 88-331 (Washington, DC, 1988).

30Data derived from National Science Foundation, *Federal R&D Funding by Budget Function, Fiscal Years 1989-91*, NSF 90-311 (Washington, DC, 1990). Defense R&D spending includes both Department of Defense (DoD) and Department of Energy atomic defense program R&D support. The latter accounts for about 7% of the total R&D budget devoted to national defense. All data are shown in actual budget authority.

31 Not including DoE atomic defense program.

32 Nonprofit institutions get another 1%. Data source is National Science Foundation, *Federal Funds for R&D: Fiscal Years 1988, 1989, and 1990*, XXXVIII, NSF 90-306, Detailed Historical Tables (Washington, DC, 1990).

33 In each of these categories government laboratories perform from 25% to 40% of the total R&D. The development portion of the military R&D budget grew significantly over the 1980s decade. In 1980 it was only 84%, whereas in 1989 it was 91%. The biggest loser was applied research, which dropped from 12% in 1980. Data derived from National Science Board, *Science & Engineering Indicators—1989,* NSB 89-1 (Washington, DC: U.S. Govt. Printing Office, 1989), Appendix table 4-6, 272ff.

34 Some DoD agencies, especially DARPA, consider programs budgeted as applied to contain a significant portion of work that may more properly be considered basic research. However, this basic research is funded primarily because of its potential value to promoting long-term applied technology

ends.   Hence, military-sponsored basic research has often been called "mission-directed" basic research.   For discussion of the blurring of basic and applied research boundaries in DoD programs, see Joel S. Yudken and Barbara B. Simons, *A Field in Transition, Federal Funding of Academic Computer Science—Current Trends and Issues.* Final Report of the Project on Funding Policy in Computer Science, SIGACT, Association for Computing Machinery (forthcoming).

[35] This is based on figures for federal and DoD support for industrial R&D in National Science Board, *Science & Engineering Indicators—1989*, Appendix table 6-2, p.351 and p.94.   About 28% of all funding for industrial R&D comes directly from the DoD.   If unreimbursed funding for independent research and development (IR&D)—research tied to potential military projects funded out of afirm's own pockets—this figure goes over 30%.   If DoE and NASA industrial R&D funding is added, then 36% of all industrial R&D is military-space-energy-related.

[36] David Henry and Richard Oliver, "The Defense Buildup, 1977-85: Effects on Production and Employment." *Monthly Labor Review* (August 1987): 3-11.   Henry calculated the direct and indirect purchases attributable to Pentagon orders for all American industries.   His procedure enables us to track the rounds of indirect suppliers, improving upon estimates based on prime contracts alone.   Henry did the industry estimates, while Oliver contributed those for employment, based on Henry's industry totals.   See also David Henry, "Defense Spending: A Growth Market for Industry." *U. S. Industrial Outlook,* 1983, XXXIX-XLVII.

[37] NASA prime contracts have been in the range of 5 to 10% of DoD budgets.   Sales of military aircraft to foreign governments were 22% of sales to the Pentagon in 1987.   Some measure of the underestimation here can be gauged by looking at the category "missiles," where there has been literally no non-military demand over the period—yet it registers only 78% defense-dependent with the DOC data.   A conservative adjustment upward of these figures to account for such sales would result in military dependency ratios of 100% for missiles, 79% for aircraft, and 60% for communications equipment.

[38] A third set of industries is beneficiary of military spending in rather large absolute amounts without being dependent on them.   Among them are computers and semiconductors, the two miracle industries spawned by military R&D in earlier decades, and several materials (steel, chemicals, oil) and service industries (airlines).   Together with others not listed like railroads, real estate and wholesaling, these defense-benefiting industries sold $35 billion (in 1977 dollars) worth of goods and services to the Pentagon in 1985.   For further discussion see Ann Markusen and Joel Yudken, *Dismantling the Cold War Economy* (New York: Basic Books, 1992).

39 Murray Weidenbaum, "The Transferability of Defense Industry Resources to Civilian Uses," in Convertability of Space and Defense Resources to Civilian Needs: A Search for New Employment Potentials, Subcommittee on Employment and Manpower, 89th Congress, 1964, reprinted in James Clayton, (Ed.), *The Economic Impact of the Cold War* (New York: Harcourt, Brace & World, 1970), p. 98.

40 Aerospace Industries Association of America, *Aerospace Facts and Figures, 1989/1990,* (Washington, D. C.: AIA, 1989), pp. 104-105.

41 National Science Foundation, *Science and Engineering Personnel,* 6.

42 Henry and Oliver, "The Defense Buildup, 1977-85," 10.

43 National Science Foundation, *Science and Engineering Personnel,* 14.

44 Henry and Oliver, "The Defense Buildup, 1977-85," 7. We believe these to be low estimates. See Ann Markusen and Scott Campbell, "Defense Spending and the Occupational, Industrial and Regional Distribution of Economic Activity," Paper presented at the Regional Science Association meetings, Boston, MA, November 1990.

45 Almost all of the net increase came in durable goods industries. Henry and Oliver, "The Defense Buildup, 1977-85," 8.

46 For further discussion of this point see Ann Markusen and Joel Yudken, *Dismantling the Cold War Economy*, Chapter 6.

47 Ibid.

48 National Science Foundation, *Science and Engineering Personnel,*14.

49 Ibid., 10.

50 Ibid., 12.

51 See Lloyd J. Dumas, "University Research, Industrial Innovation, and the Pentagon," in John Tirman, ed., *The Militarization of High Technology* (Cambridge, MA: Ballinger, 1984), pp. 138-139; Warren F. Davis, "On the Number of Engineers and Scientists Serving the Defense Sector." Paper presented to "The War Against the Economy: The Military-Industrial Complex," a colloquium sponsored by the Columbia University Seminar on the Political Economy of War and Peace, Columbia University, January 26, 1985 (unpublished). Davis, who conducted his own survey of university placement officers and industry campus recruiters, supports Dumas' claim of 30%, stating that his "survey results indicate that approximately 27.8% of those graduating annually from engineering and science departments across the country enter defense employment."

52 A similar unsubstantiated figure of 20%-30% of engineers working in defense industries was also stated in an otherwise antidepletionist article in *Science* . Eliot Marshall, "The Pentagon Is Not Eating Engineers," *Science* 234 (November 1986): 664-665.

53 National Science Foundation, "Projected Response of the Science, Engineering, and Technical Labor Market to Defense and Nondefense

Needs: 1982-87", Special Report NSF 84-304, Washington, DC, January 1984, vi. Quoted in Warren F. Davis, "On the Number of Engineers and Scientists Serving the Defense Sector."

54 U.S. Department of Defense, "Estimates of Industrial Employment by Occupation (Engineers & Scientists)," Defense Economic Impact Modeling System (Occupation by Industry Model) (Washington, DC: Government Printing Office, 1983).   Cited in Lloyd J. Dumas, "University Research, Industrial Innovation," 138.

55 DoE-funded researchers, however, will likely have a more direct awareness if their work is defense-connected or not than those supported by NASA, especially those directly engaged in DoE's atomic defense program. There are some, however, in the energy sciences area who can be considered "dual-use" and might therefore be legitimately counted as defense-related.

56 Warren F. Davis, "On the Number of Engineers and Scientists Serving the Defense Sector."

57 Murray Weidenbaum,"The Impact of Military Procurement on American Industry," in J. A. Stockfisch (ed.), *Planning and Forecasting in the Defense Industries* (Belmont, CA: Wadsworth, 1962), pp. 135-174; William Baldwin, *The Structure of the Defense Market, 1955-1964* (Durham, NC: Duke University Press, 1967), 84-85.

58 Herman O. Stekler, *The Structure and Performance of the Aerospace Industry* (Berkeley: University of California Press, 1965):   100.

59 Aerospace Industries Association of America, *Aerospace Facts and Figures, 1989/1990*, (Washington, DC: AIA, 1989): 104, derived from Battelle data.

60 Seymour Melman, *The Permanent War Economy* (New York: Simon & Schuster, Inc., A Touchstone Book, 1985), p. 244.

61 "Lockheed to Hire 2600 in 1984," *Oakland Tribune,* 31 January 1984: 1.

62 The mix of blue collar and S&Es also varies with the production cycle. At Textron's Avco Company, on Route 128 outside Boston, the movement from the design phase into the production of a missile boosted the blue collar share of total firm employment from 25% to 50% in just two years. Jacques Gansler, *The Defense Industry,* 192, and interviews with Hughes and Avco.

63 The Bureau of Labor Statistics and the Department of Commerce examined in a joint study the economy-wide defense dependency of industries and occupations. Henry and Oliver, "The Defense Buildup, 1977-85," 3-11.   Henry and Oliver took federal budget outlays (not appropriations), converted them to calendar year figures and deflated them to constant 1977 dollars. These final demand totals were then run through the DOC's updated 1977 input-output model to estimate the shares of industrial

output directly and indirectly associated with defense procurement for 1985 and earlier years. With Department of Labor estimates of occupational demand per unit of output for each industry, employment associated with defense spending in each industry was then estimated. Requirements by industry and occupation are drawn from detailed occupational and industrial surveys, updated with nonsurvey techniques. The only alternative to this data base is the DEIMS model, built by an outside contractor for the DoD, which Henry claims is not as accurate as the DOC's model because it does not use as much original and updated survey data. Ann Markusen and Scott Campbell worked Henry and Oliver's unpublished data, subsequently generously supplied to the former analysts, to produce defense-dependency estimates by industry and occupation.   A detailed presentation of their methodology and findings are reported in   Ann Markusen and Scott Campbell, "Occupational, Industrial and Regional Distribution of Defense-Related Economic Activity." See also Ann Markusen and Scott Campbell, "Estimating Defense-Dependent Output and Employment by Industry and Occupation," Technical Note #1, Project on Regional Labor Pool Formation and Occupational Structure in U.S. Defense-Oriented Industries, Project on Regional and Industrial Economics, Rutgers University, May 1990.

[64] The SSE is a panel study conducted by the National Science Foundation. It tracks a sample of scientists and engineers drawn from the 1980 Census.   Unlike the BLS data, which are based on workplace surveys, the SSE data are based on a sample drawn from a survey of households and resurveyed every two years. Entrants to the labor force since 1980, predominantly young people, are excluded, and by 1986 they may comprise as much as 30% of the occupational group. Because they probably enter the defense side of the business in disproportionately large numbers, their exclusion will result, by 1986, in underestimation of defense-related occupational shares.

[65] For fuller discussion of the BLS methodology and of the SSE and its methodological problems, see Markusen and Campbell, "Occupational, Industrial and Regional Distribution of Defense-Related Economic Activity." They note, for example, that the SSE figures do not include new entrants to the labor force since 1980, which excludes around 30% of the 1986 science and engineering work force. "If anything," they state, "we think this yields a downward bias in the SSE estimates."

[66] This leaves us with a rough estimate of the defense dependency of S&E occupations that are most closely associated with military R&D and production. Another reasonable deduction is the large bloc of computer specialists whose main function is to provide computer application services for businesses, government bureaucracies, and other institutions. In our opinion these jobs should not necessarily be lumped together with computer scientists and engineers whose functions are related more to the research,

design, and development of computer technologies. If we deduct computer specialists who reported their primary work as reporting/statistical and computing tasks from the NSF S&E employment data base but assume that the number of defense-dependent S&Es in 1986 is not affected (which introduces some distortion), then the approximate defense dependency ratio for computer scientists rises to 21%, for all scientists it grows to more than 18%, and for the total defense-relevant S&E work force, it increases to 20%. Data derived from National Science Foundation, *Science and Engineering Personnel;* and Markusen and Campbell, "Occupational, Industrial and Regional Distribution."

[67]Ann Markusen, Peter Hall, and Amy Glasmeier, High Tech America (Boston: Unwin & Hyman, 1986): Table 2.5.   Based on the 1980 Occupational Employment Survey, United States Department of Labor.

[68]Department of Defense, "500 Contractors Receiving the Largest Dollar Volume of Prime Contract Awards for RDT&E," Annual.

[69]Compiled by Ann Markusen from unpublished data from the National Science Foundation's Survey of Scientists and Engineers (SSE) and the Bureau of Labor Statistics (BLS).

[70] For a fuller discussion of these regional variations see Markusen and Campbell, "Occupational, Industrial and Regional Distribution."

[71] These findings from Markusen's and Campbells work on the SSE corroborate findings from earlier decades. In the 1960s, for instance, a survey of engineering salaries showed that for an engineer with 10 years of experience, machinery-producing industries paid $ 9300 on average while the aerospace industries paid $ 11,500. See Seymour Melman, *Our Depleted Society,* reprinted in James Clayton (ed.), *The Economic Impact of the Cold War* (New York: Harcourt, Brace and World, 1970).

[72] Research funding refers only to basic and applied research. Total federal spending on R&D in academia in fiscal 1988 was $7.8 billion, of which $4.9 billion, or 62%, went for basic research and $2.1 billion, or 28%, was for applied research, and only 10% was for development. Data derived from National Science Foundation, *Federal Funds for R&D: Fiscal Years 1988, 1989, and 1990.*

[73] It should also be noted that certain subfields within these disciplines may be particularly defense-dependent. Within computer science, for example, artificial intelligence and computer architectures get almost all their funding from the DoD, while theoretical computer science depends almost exclusively on the National Science Foundation for its support. For discussion of the DoD role in computer science, see Joel S. Yudken and Barbara B. Simons, "Federal Funding in Computer Science: A Preliminary Report," *A Field in Transition SIGACT News* 19, No.1 (Fall 1987).

[74] National Science Board, *Science and Engineering Indicators—1989,* 10.

75 This is a significantly greater concentration than other agencies, such as the NSF. See Yudken and Simons, *A Field in Transition*.

76 Data derived from National Science Foundation, U.S. Scientists and Engineers: 1988; National Science Foundation, *Characteristics of Doctoral Scientists and Engineers in the United States: 1987*.

77 It isn't clear from the data sources whether the figures on academic S&Es cover heavily DoD-funded academic-affiliated research institutes such as Johns Hopkins' Applied Physics Lab or USC's Information Sciences Institute. The data also do not include academically affiliated researchers who work at, have joint appointments with, or consult with DoD-funded FFRDCs such as Carnegie Mellon's Software Engineering Institute or MIT's Lincoln Labs, or DoE's national weapons laboratories at Livermore and Los Alamos. S&Es, particularly those who hold doctorates, are employed in large numbers at these institutions and at nonprofit institutes such as SRI International and Charles Draper Labs, which also receive large DoD contracts. None of the available data sources seem to fully cover these important classes of military-related S&Es.

78 National Science Foundation, *Science and Engineering Personnel*, 18.

79 These figures by agency are derived from National Science Foundation, *Federal Scientists and Engineers: 1986*, Detailed Statistical Tables, NSF 87-320 (Washington, DC, 1987). These tables treat computer scientists as a separate category rather than as a field of science, as the other sources used in this paper do. The numbers reported here are somewhat lower and in different proportions than reported in National Science Foundation, *Science and Engineering Personnel*. The reason for the discrepancies is that the reports are derived from different data bases. The figures presented here for federal S&E employment by agency should therefore be treated as only rough approximations. According to these data, 37% of the federal S&E work force were scientists, 44% were engineers, and 18% were computer scientists in 1986.

80 The federal S&E data refer only to civilian employees.

81 Jacques Gansler, *Affording Defense* (Cambridge, MA: The MIT Press, 1989), p. 248.

82 Mary Kaldor, *The Baroque Arsenal* (New York: Hill and Wang, 1981), p. 192.

83 R. P. Loomba, "A Study of the Re-employment and Unemployment Experiences of Scientists and Engineers Laid off from 62 Aerospace and Electronics Firms in the San Francisco Bay Area During 1963-65" (San Jose State College, Manpower Research Group, Center for Interdisciplinary Studies, 1967).

84 Trevor Bain, "Labor Market Experience for Engineers During Periods of Changing Demand," Final Report, (Washington, DC: Office of Research

and Development, Manpower Administration, Department of Labor, 1973): 1.

[85] Cited in Rick Wartzman, "Defense Contractors Gird for Warming of the Cold War," *Wall Street Journal* (11 November 1989): A8.

[86] Linda Kravitz, "The Wages of Peace," Review draft, Summary and Findings, (Washington, DC, Office of Technology Assessment: August 1990): 10.

[87] Russell Mitchell, "Instead of Star Wars, Little Star Wars?" *Business Week*, (1 July 1991): 80-81.

[88] See Defense Budget Project, *Responding to Changing Threats,* A Report of the Defense Budget Project's Task Force on the FY 1992-FY 1997 Defense Plan (Washington, DC, June 1991).

[89] National Science Foundation, Federal Funds for Research and Development, Federal Obligations for Research by Agency and Detailed Field of Science/Engineering, Fiscal Years 1969-1990, Detailed Statistical Tables (Washington, DC: Division of Science Resources Studies).

[90] National Science Foundation, *U.S. Scientists and Engineers: 1988, Estimates,* Detailed Statistical Tables, NSF 88-322 (Washington, DC, 1988).

[91] National Science Foundation, *Characteristics of Doctoral Scientists and Engineers in the United States: 1987,* Detailed Statistical Tables, *NSF 88-331* (Washington, DC, 1988).

[92] National Science Foundation, *Science and Engineering Personnel,* 32. These and most of the other supply-demand data used in the following text comes from this report.

[93] Ibid., 35.

[94] Ibid. Between 1988 and 2000 this sector is projected to provide approximately 550,000 jobs for scientists and 600,000 for engineers.

[95] Ibid., 29.

[96] Ibid., 30.

[97] Ibid., 39.

[98] Ibid., 42.

[99] Ibid., 29.

[100] Ibid., 32.

[101] Ibid., 48.

[102] Robert Buderi, "The Brakes Go on in R&D," *Business Week* (1 July 1991): 24-26.

[103] The rate of shifting into defense work was slightly higher—27%—on the part of those who had been in civilian jobs in 1982. Lerner, looking at a subsample of the SSE, found that 24% of those supported by defense funding in 1982 reportedly shifted to nondefense sponsorship in 1986. Furthermore, of those working with defense funding in 1986, 26% had not had such support in 1986. Joshua Lerner, "The Mobility of Corporate Scientists and

Engineers between Civilian and Defense Activities: Evidence from the SSE Database." Discussion Paper 90-02, Science, Technology and Public Policy Program, John F. Kennedy School of Government, Harvard University, August 1990.

104 Ibid., 22.

105 Seymour Melman, *The Permanent War Economy*, 234.

106 Carl Rittenhouse, *The Transferability and Retraining of Defense Engineers*, Stanford Research Institute, November, 1967, 41. Cited in Robert DeGrasse, Alan Bernstein, David McFadden, Randy Schutt, Natalie Shiras, and Emerson Street, *Creating Solar Jobs, Options for Military Workers and Communities*. A Report of the Mid-Peninsula Conversion Project (Mountain View, CA: Mid-Peninsula Conversion Project (now Center for Economic Conversion, November 1978): 30.

107 The examples below are taken from Ann Markusen and Joel Yudken, *Dismantling the Cold War Economy*.

108 Panel on Engineering Labor Markets, Office of Scientific and Engineering Personnel, National Research Council. *The Impact of Defense Spending on Nondefense Engineering Labor Markets* (Washington DC: National Academy Press, 1986): 81-82; Linda Levine, "Defense Spending Cuts and Employment Adjustments," *CRS Report for Congress # 90-55E* (Washington, DC: Congressional Research Service, Library of Congress, June 27, 1990): 9.

109 Marion Anderson, Converting the Work Force: Where the Jobs Would Be (Lansing, MI: Employment Research Associates, 1983), p.6.

110 Linda Levine, "Defense Spending Cuts and Employment Adjustments," 9.

111 Several analysts have noted that there are really two separate labor markets that defense engineers face—one for new graduates and one for experienced workers. See Trevor Bain, "Labor Market Experience for Engineers"; and Max Rutzick,"Skills and Location of Defense-Related Workers." *Monthly Labor Review* 93 (February 1970): 11.

112 "Who Pays for Peace?" *Business Week* (2 July 1990): 67.

113 Stephen Engelberg, "Pentagon Imposes a Hiring Freeze that Could Eliminate 50,000 Jobs," *New York Times* (13 January 1990): 1-11.

114 See Hilary Wainwright and Dave Elliott, *The Lucas Plan, A New Trade Unionism in the Making?* (London: Allison and Busby, Ltd., 1982); Dave Elliott and Hilary Wainwright, "The Lucas Plan: The Roots of the Movement," in Suzanne Gordon and Dave McFadden (eds), *Economic Conversion, Revitalizing America's Economy* (Cambridge, MA: Ballinger Publishing Co., 1984): 89-107. For more on Western European experiences in the early 1980s, see also Suzanne Gordon, "Economic Conversion Activity in Western Europe," in Gordon and McFadden (eds), 108-129. For a recent review of early eighties bottom-up conversion efforts see also

Catherine Hill, Sabina Deitrick, and Ann Markusen, "Converting the Military Industrial Economy: The Experience of Six Facilities," *Journal of Planning Education & Research,* 1992.

115 For discussion of the McDonnell Douglas effort see Joel S. Yudken, "Conversion in the Aerospace Industry: The McDonnell Douglas Project," in Gordon and McFadden (eds), *Economic Conversion,* 130-146. See also Hill, Deitrick, and Markusen, "Converting the Military Industrial Economy."

116 This and the following examples are discussed in detail in Robert W. DeGrasse Jr., with Anna Seiss, "Previous Industrial Conversion Experiences," (Arlington, VA: The Orkand Corporation, 1985). This report was substantially incorporated into President's Economic Committee and the Office of Economic Adjustment, Office of the Assistant Secretary of Defense (Manpower, Installations and Logistics, Department of Defense, *Economic Adjustment/Conversion, Appendices* (Washington, DC:The Pentagon, July 1985): Appendix M.

117 Robert DeGrasse, Jr., "Previous Industrial Conversion Experiences," 13-17.

118 Ibid., 20.

119 "How Rohr's Move into Transportation Backfired," *Business Week* (19 January 1976): 48. Quoted in Robert DeGrasse, Ibid., 25. DeGrasse also observes that "Rohr learned that producing a new generation of railcars required a much longer development process than the firm had originally expected."

120 Ibid., 21.

121 Carl Rittenhouse, *The Transferability and Retraining of Defense Engineers*: 8.

122 See Ann Markusen and Joel Yudken, *Dismantling the Cold War Economy*: especially Chapter 9. See also Joel S. Yudken and Michael Black, "Targeting National Needs: A New Direction for Science and Technology Policy, *World Policy Journal* (Spring, 1990): 251-288.

8
The Engineering Profession and Local Economic Adjustment

Kenneth Matzkin

Abstract

The Office of Economic Adjustment (OEA) of the Department of Defense (DoD) provides financing, planning expertise, and coordination for communities impacted by base closings and cancellations of DoD contracts. Established in 1961, this office has the greatest experience with the planning associated with economic conversion. OEA uses the technical skills of its permanent staff, contracted consulting services, and planning grants to address the problems of affected communities.

OEA targets its assistance to the community level. Planning efforts help to organize communities and to identify solutions for the loss of economic activity associated with base closings or the loss of contracts. Since DoD employs a large number of engineers at bases and through defense contractors, the communities affected by defense cuts may include large numbers of engineers. These engineers should be involved in a community's planning efforts through the use of their technical skills. Engineering resources could be used to assess the viability of industries that are identified by community planners as replacements for DoD's presence.

The Defense Economic Adjustment Program

The Office of Economic Adjustment provides technical assistance to communities through its permanent staff, contracted consulting services, and planning grants through programs offered to members of the President's Economic Adjustment Committee (EAC). EAC is an interagency organization established and maintained under Presidential Executive Order (Exec. Order 12788; Jan. 15, 1992). It coordinates and directs the normal program aid of its twenty-three federal member agencies to communities affected by the Department of Defense (DoD). This approach is intended to assure implementation of communities' plans to adjust to DoD-induced changes.

All informed projections show that DoD will reduce its activity during the 1990s. A reduction of 25% is projected by 1995. This reduction will include closing installations, reducing the number of uniformed and civil service personnel, and buying less equipment and weapons. These projections dictate that an increasing number of communities will experience socioeconomic upheavals as a result. Some may be severe.

Community Adjustment Process

The OEA aims its assistance at geographic regions, particularly communities. For the OEA, a community is a network of people and their institutions: workers, students, elected representatives, businesses, schools, hospitals, and so on. OEA's experience shows that a successful way to mitigate community-wide impacts is through an integrated, coordinated, grass roots effort aimed at creating a consensus of what should be done to offset the DoD-induced negative impacts. This consensus is often reached by recognizing the uses to which a closed DoD installation can be put. The DoD monitors how almost one hundred former military installations are now being used. Their investigation, *The Civilian Reuse of Former Military Bases* reports that 93,424 civilian jobs have been replaced by 137,823 new jobs at these sites.[1] Thus, assembling representatives of a community and pointing out possibilities comprise the essential first step to overcoming the loss of funds flowing into a community from military cuts.

The resulting organization seeks to discover various institutional perspectives on the problem. For example, a local public school district will have to prepare for fewer students and smaller DoD subsidies. By contrast, a local construction industry will see new construction and rehabilitation opportunities when the base is later reused. Identifying, accounting for, and coordinating legitimate but possibly conflicting points of view and consequent psychological reactions are essential before feasible mitigation plans can be devised.

The OEA has found that solutions to the loss of economic activity in the community are identified most readily through a formal planning program instituted by the community organization. The community itself must be analyzed, its strengths and weaknesses as a place to live and work must be identified, and recommendations about the pursuit of compatible economic activities must be discussed. Industrial development experts can help this discussion to work, but these assessments must include intensive, guided participation by community representatives.[2] If an asset such as an abandoned DoD or private enterprise's installation/facility becomes available, it must be assessed as a possible vehicle for generating economic activity. An OEA publication *Diversifying Defense Affected Economics* reviews five examples of how communities diversified their economic activities using shutdown DoD resources and labor.

The planning cycle also identifies actions the community can take. The resulting plans are then translated into a variety of subsidiary actions to enable economic activity: (1) jobs and tax-producing enterprises must be identified; (2) public infrastructure (e.g., water and transportation networks) sufficient to meet those enterprises' needs must be constructed; and (3) land use regulations must be altered to ensure compatibility among various commercial, industrial, residential, and public owners and users of land. As with any human endeavor, successful community and economic development can later require expanded and even new public services in such diverse areas as public safety, elementary and advanced education, and environmental protection.

Mitigation/Development Strategies

Although the key to mitigating local economic distress is replacement/expansion of local opportunities, communities are more than aggregates of jobs. Successful community planning is enhanced if the quality of life, geographic and environmental attributes, business climate, demographic trends, educational system and characteristics, social/charitable institutions, political history, and other aspects that affect a person's decision to live, work, and invest in that community are taken into account.

Most development strategies first look at local employment opportunities already in place. Their attention can be turned to creating, expanding, or attracting new economic activities (i.e., employers) best suited to enhance a community's specific attributes: leadership structure, geography and climate, labor force, public support services, etc. Whether to pursue public- or private-sector employment generators should also be taken into account. For example, does the community fit the Federal Bureau of Prison's profile to host a correctional center (a remarkably powerful economic activity); or is the community underserved, by comparison with other similar places, by a key retail sector that could be introduced and thrive there? These are complex issues that must be assessed using empirical data that are tempered by the community's "vision" of how it wants to be.

For the impact adjustment plan to be successful, a broad base of the people in the community must participate. Many of the implementation projects—as disparate as zoning changes, capital enhancement programs, skills training courses, and old-fashioned "grantsmanship"—will require expertise and other resources not always embodied in the organizing and planning team.   If the plan is artfully introduced to the people and institutions of the community who can implement it, the plan has a greater possibility of success.

Roles for Engineers

Large numbers of engineers (and scientists) are expected to lose jobs with DoD contractors.   Although distressing, their availability for alternative career directions can provide an important dimension to the community's adjustment planning effort.   Staying informed about probable trends among engineers soon to be dislocated will better enable a community to select and define its adjustment strategies.   Planners will need to know what engineers would most like to do: start their own (R&D, manufacturing, marketing, or consulting) businesses, pursue academic careers, buy out or join a local business, or relocate to another community.

Engineers can contribute to impact adjustment plans in several ways.   By assessing industries recommended, the community engineers can evaluate which technologies promise long-term viability and jobs.   At the same time, they can also recommend emerging, promising technologies (and related industries) that have not yet been discovered by less technically current local firms.

In several communities where dislocations are occurring, groups of engineers have established mutual support networks and have even secured facilities to accommodate them.   One such program is the Center for Practical Solutions located in Hauppauge, New York.   These networks provide services ranging from crisis-intervention counseling to trading leads on employment.   They also offer a venue for engineers to brainstorm about technology applications, possible inventions, the demand for individual, and even ad hoc joint consulting services, and other related productive initiatives.

Another current trend in American industry and government is to use cooperative labor-management committees to address issues that will keep the organization vital.   By looking at improving labor skills, choosing and using capital equipment, applying new technologies and developing products, such committees improve performance and prepare an enterprise for negative outside influences.   Such committees also spot internal de-stabilization.

Conclusion

Engineers are not often community activists but work well as a team. To adjust, communities need the fullest range of ideas, energy, and commitment possible. And engineers bring a perspective and knowledge that may assist economically weakened communities in selecting alternative industries. The task of converting job opportunities from the declining DoD marketplace requires creativity and problem-solving skills and should be looked at as a positive challenge. Isn't that the essence of engineering?

Notes

1 Office of Economic Adjustment, *The Civilian Reuse of Former Military Bases* (Washington, DC: Office of Economic Adjustment, 1990).

2 Office of Economic Adjustment, *Diversifying Defense Affected Economics* (Washington, DC: Office of Economic Adjustment, 1991).

9

The Sociology of the Engineering Profession: Engineers and Economic Conversion

Evan Vlachos

Abstract

This chapter addresses the following questions.

How will the technical community be affected? If economic conversion is unplanned, the technical community will likely experience significant dislocations and associative stresses and strains in their communities. After being laid off, many defense-sector engineers, especially older engineers, have difficulty finding adequate jobs in well-established, desirable locales. Job dislocations and employment changes will probably drop salaries. On the other hand, engineers who can develop new models and paradigms of thinking, new designs, and ways to implement designs, can overcome their narrow specialization and may find numerous opportunities in economic conversion.

How is the technical community likely to affect economic conversion? As part of a broader ongoing socioeconomic transformation, the technical community can react positively by facilitating the transition from defense to nondefense work. The capacity for team work and highly specialized products where the U.S. has been competitive, such as aerospace and electronics, offer promising areas into which engineering talent could be directed.

How can the technical community participate constructively in the conversion? Engineers who want to foster conversion can participate actively and constructively in joint ventures between defense contractors, universities, and corporations and build strategic alliances. Engineers may be more likely to support conversion if they understand the nature and character of national and international economies and can participate in continuing education to reorient their careers and upgrade their skills. Professional engineering societies can take a more aggressive stand in promoting public policy involvement, by focusing attention on macro-engineering projects in environmental challenges and infrastructure maintenance and rehabilitation.

What are the key gaps in our knowledge? Research is needed on the composition of the current engineering labor force and projections for future engineers: What are the characteristics of those working in defense as compared to civilian engineers? How does the composition of the present and future engineering work force (especially women and foreign-born) affect the conversion process? Can we have better estimates of the dependence of the engineering profession on military-oriented activities? Will there be an engineer shortage and in what form? What is the curriculum needed to develop an adaptable, knowledgeable, problem-oriented engineer? What considerations and sources of information do engineers use in making education, training, reorientation, and other career choices?

This paper uses the metaphor "from swords to ploughshares" to develop sociological dimensions of conversion from a military to a more civilian oriented engineering practice and economy by exploring: (1) the history of the engineering practice and its standing in today's complex society; (2) an outline of the professional framework and shifting worldviews; (3) the engineering standing in the community and societal engagement; and (4) prospects and options associated with the conversion process.

Background

Engineering in a Complex Society

Many observers perceive the end of the twentieth century as both a catastrophic threat and an opportunity in which we find passage to the new era of the information society. The momentous events of the last few years have been variably described as the end of history, the end of the Cold War, the coming of a new order, the globalization of human affairs, or the coming of the post-industrial/cybernetic society.

All such terminologies and preoccupations provide the impetus for an understanding of what might be called a "Grand Transformation" in economic, political, technical, cultural, and social terms. Central to such far-reaching changes are the driving forces of technology and accelerating sociopolitical transformations.

From a sociological point of view, conversion involves not only the physical and economic transformation of plants and equipment but also the changing of values, attitudes, skills, and the entire cultural outlook of the work force. The process of economic conversion can be seen as concentric circles starting from the individual and expanding to the community, organizational, national, and even the global environment. This economic conversion and its broader societal transformation imply the changing of personal outlook, the building of new stakeholder groups, the writing of new laws and guidelines, the definition and prioritization of new national and global alternatives, the retraining and retooling of the work force as well as anticipatory and participatory planning in defining goals and objectives of communities and societies.

The redirection of labor and capital resources has been accelerated with recent political developments including the end of the Cold War, the realignments in Eastern Europe and the Soviet Union, and the subsequent reallocations of natural and economic resources. More recently in the United States, discussions of conversion have focused on the trauma of the closing of the military bases and provided a vivid example of the need for streamlining the economy, a more efficacious and equitable reorientation of social priorities, mobilizing personnel and resources in engineering projects relevant to massive social problems, redirecting training and academic emphasis, utilizing the technological prowess seen in recent "smart weapons" for the production of civilian goods, and finally the renovation of infrastructure and the promotion of community welfare.

Conversion highlights: (1) new areas of concern and policy emphases for a work force where one-third of all engineers involved in military projects and 80% of the federal R&D is defense-related, (2) a transition to a new economy characterized by global competitiveness and increased emphasis on productivity, and (3) the search for new values and skills to make the transition to a post-industrial cybernetic society.

In this context, the shocks of the 1990s provide for an interesting parallel with the period around 1964, which produced some pronounced shocks for the complacent engineer. Events at that time (the nuclear test ban treaty, a declining defense budget, sharp restrictions on new funding for strategic forces, the phasing out of many defense installations, and the cancellation of contracts) provide striking parallels with current discussions as to ongoing or desired transformations. Studies published at that time documented the experiences of engineers and scientists who had to move from an occupational subculture geared to military emphasis to new industries that were nondefense-oriented.[1]

The consequences of the 1964 defense shrinking, layoffs, subsequent recessions, and, more recently, the 1991 preoccupation with a new world economic order can no longer be viewed as periodic cycles that affect only particular professions. Rather they are the products of complex socioeconomic problems that call attention to the character and nature of government spending and the types of employment and levels of affluence. Other factors include the self-image of engineers and other professionals and a new socioeconomic order characterized by constraints, competing and conflicting demands, global interdependence, worldwide competition, and shifting priorities and preferences. This new socioeconomic order emphasizes "positive economic nationalism," where each nation's citizens take primary responsibility for enhancing the capabilities of all citizens for a full and productive life while recognizing at the same time the existence and needs of other nations.[2] The challenge of this transformation is driven by expected scale breakdowns of complex systems manifested as environmental problems, climatic changes, and institutional change from an industrial to a post-industrial society. What is needed is an emphasis on both short-term coping and long-term adaptation to a transforming environment where the key element in terms of professional development is how to recruit, retain, and renew the engineering work force of the nation.

In any discussion of this transformation one can illustrate present concerns and viewpoints in three driving metaphors. The first is the traditional expression from "swords to ploughshares," or the transformation from a military to a civilian economy and culture.[3] The second can be described as "the moon and the ghetto" metaphor, which emphasizes how scientific knowledge and technological capabilities have evolved in this country.[4] Simply expressed, it implies that if we have the technical capacity to send someone to the moon, then why can't we solve social problems? This metaphor is misleading as a policy analogy because the complexity of social life makes it difficult to apply such focused resources and activities across a wide range of social and economic problems.

Finally "the end of history" is the third metaphor, recently used to describe the end of a grand ideological battle between free market economies and centralized, directed economies.[5] However, more recently ethnic, religious, and subcultural confrontations and outright militant conflicts have dampened the euphoria about a war-free world and inserted caution as to the obsolescence of traditional military preparedness.

The New Context

Central to the conversion argument is the preoccupation with an overburdened economy divided into contributive, neutral, and distractive sectors.[6] Such categorization implies that economic activities are not the ultimate good: beyond economic priorities greater good is the quality of human life. Therefore, any economic activity that produces the necessaries and conveniences of life is of itself merely an instrumental goal to be evaluated in terms of how it affects the quality of life and social well-being. The preoccupation with an inefficient or overburdened economy points out that the effectiveness of a desired transition should be judged by the extent to which costs are minimized and by the rate and success of the transfer of economic activity to the contributive sector.[7]

This raises questions as to how, on the one hand, the technical community will be affected by the far-reaching institutional shift of conversion; and, on the other, how the technical community is likely to affect institutional priorities. The transfer of engineers and scientists from distractive-sector (military) activities to the contributive sector (civilian) highlights differences in the development and application of technologies. The high-tech military products of today are extremely complex, designed to squeeze every possible source of performance out of a product. This leads to narrow specialization of engineers.[8] The traditional civilian market, on the other hand, requires much broader knowledge, sensitivity to social demands and, consequently, retraining and reorientation for military-oriented scientists and engineers. Furthermore, in order for engineers to participate constructively and in a participatory and anticipatory fashion in economic conversion, we need to shift the curriculum of universities to include different emphases in courses taught as well as epistemological approaches in training, research, and implementation.

But such a challenge should not daunt us, as we have faced other similar large-scale challenges. After the end of World War II, the United States economy went through a remarkably successful large-scale transition from military to civilian-oriented production.[9] It is possible to redirect a heavily military-oriented national economy without extraordinary disruption to the life of the nation. Yet, although the past points to our ability to convert defense industries following World War II and the Korean and the Vietnam wars, some major differences mark our era. The general level of unemployment today is much higher than previous post-war periods. At the same time, the military market and military technology have become a global, multinational industry whose specialization inhibits finding possible civil uses for technology developed in weapons production programs.[10] This raises the issue of transferability; i.e., what are the spin-offs and the spillovers of military and space hardware to everyday consumer life?

Engineers as a Work Force

Current geopolitical events challenge the role of engineering within our society and our reliance on this profession for national policy making and fiscal and long-term socioeconomic planning. The question is the extent to which scientists and engineers by tradition, professional ethos, or occupational composition are capable of and willing to contribute to economic conversion and to a better understanding of how the technology development works in a new context.[11]

In order to understand the potential role of engineers in conversion, we must first understand the engineering infrastructure and the definition of an engineer, engineering, and the engineering community. The traditional focus on the engineer has been replaced by a focus on the engineering community in recognition of the distinctive culture that is more connected than a mere aggregation of engineers.

In the late 18th and early 19th centuries, engineers performed two types of activities in the United States. The most prominent was civil engineering, which focused on public works such as the building of canals, roads, and fortifications as well as the installation of water supply systems for cities. The second type was mechanical engineering, which involved the traditional producing and selling of certain types of metal goods.[12] The middle of the 19th century was the golden age of engineering, which created an enormous sense of professional pride, culminating with the famous statement of Samuel Smiles, who claimed grandly that "our engineers may be regarded in some measure as the makers of modern civilization."[13] The flocking of people to the great exhibitions in London, Paris, and Chicago and the writing in the literature of great travelog in the vein of Jules Verne created a culture of admiration for engineering feats. The word engineer moved beyond the military engineer to define engineering as the art of directing the great sources of power in nature for the use and convenience of people (as drafted in the Charter of the Institution of Civil Engineering in 1828 in London).

Throughout our century the nature of engineering work has been changing steadily. Now the corporate engineer has become dominant: work is characterized by large project teams, macro-engineering, relative individual anonymity, and dedication to discrete bits of technology advancement in highly specialized fields.[14] At the same time, the global sphere of corporations, the engineering community, and new technologies diversified the engineering discipline and created the major engineering disciplines (civil, mining, mechanical, petroleum, marine, agricultural, nuclear, bioengineering, engineering mechanics, environmental, ceramic, metallurgical, and materials). More important than this shift is that the engineering work force now defines itself according to the employees' functions rather than earlier training. The way employers use engineers can help us understand how easily engineers can move from one discipline to another, strategies to recruit young people to become engineers, ways to motivate those already employed, the continuing education and training needs of engineers, the effect of various initiatives on the supply and quality of engineers, and international attitudes about engineering.[15]

Rapid transformations of the workplace such as automation, robotization, digitalization, and instantization also create a great diversity of engineering-style activities within new corporate multinational frameworks. Traditional engineering is now becoming more blurred as technologists and technicians today often perform work that formerly required engineers. Conversely engineers are doing work that in the past required other professions, such as planners or managers, handling legal and social issues, and analyzing policy, political feasibility, and economic viability.

This formidable engineering infrastructure has become a barrier to conversion. Many engineers and firms have a stake in the present military establishment, the Department of Defense and its military bases, its civilian employees here and abroad, and a variety of other business firms under contractual obligations. The engineering infrastructure and the existing economy produce military and industrial sectors that are closely tied by links of mutual dependence, reward, and control. This network of federal government contracts involves about 20,000 companies and about 100,000 subcontractors. Thus every major industrial group has some stake in sales to the Pentagon.[16]

To understand the potential negative consequences of conversion on the engineering work force, it is instructive to look briefly at the supply and demand for scientists and engineers, a situation that has been described as a national crisis in the making. The demographics of the college-age population combined with estimates of those who pursue careers in science and engineering indicate significant shortfalls between supply and demand for the next several decades for both baccalaureate- and Ph.D.-level engineers.[17] The question is who will do science and engineering in the 1990s. The college-age population will drop until 1995 with some eventual demographic recovery by the year 2010, but more importantly the large proportion of foreign-born engineers from different cultural traditions who could make up for the shortage may have difficulty adjusting to the U.S. But the engineer shortage is not certain; others see no valid evidence to support the forecast of a future shortage of engineers and characterize the current situation as a period of oversupply as we reorient the economy.[18]

The labor market varies considerably among engineering specialties. New specialties such as material, bioprocess, or computer engineering often emerge as offshoots of a traditional specialty and fall between the cracks of existing data collection, making it difficult to track progress.[19] Women make up slightly more than 3% of the engineering work force and only about 3% of employed engineers hold doctorates. About 80% of science and engineering employed engineers work in business/industry[20] with the highest proportion in mechanical and electrical/electronic engineering. Such numbers coincide with the fact that the majority of the military budget goes to three key industries: aerospace, communications, and electronics.

The demographics of engineering raise questions about the implications of the influx of international students. Over 40% of recent engineering Ph.D.s have been awarded to international students on temporary visas compared to 30% in 1975. Foreign-born engineers play an increasingly significant role in American industry, but their impact is even greater in academe. The proportion of noncitizens among engineering assistant professors younger than 30 years increased from 10% in 1972 to 50%-55% in 1985.[21] Although 90% of engineering undergraduates still are U.S. born, relatively few pursue graduate studies. Foreign-born engineering faculty can influence the recruitment and retention of students insofar as cultural differences reflected in the attitudes of some foreign-born engineers lead them to discourage women and minorities from pursuing engineering studies. In addition, national security and export control regulations create barriers to the employment of foreign-born engineers in sensitive security jobs, which complicates collaborative arrangements between diverse industries and national laboratories.[22]   However, available data indicate that U.S.-born engineers have not faced appreciably diminished opportunities in industry.

Speculating further into the future, projections about the social characteristics and roles of the engineer predict that in the coming years there will be more engineers who are female, Asian, foreign-born, and from dual-career families.   Compared to the past, engineers will predominantly work in service industries, manage and communicate with people, change jobs often, undergo retraining on a sustained basis, and be more visible to the public.[23]   "Engineering's role in the future can go only one way, towards a central role in the planning and management of society."[24]

Towards a Professional Framework

Roles and Professionalization

The view that engineering is central to the life of the nation, particularly in a service and information economy operating in a more complex and globalized setting, leads to the portrait of engineers as the predestined leaders of social revolution in America.[25]   Previously, engineering progressivism provided an interesting connecting link at the end of the century between an emerging strong profession and the emphasis on technocracy and scientific management, especially in the 1930s.[26]   This engineering progressivism can be seen as part of the professional development of engineering in America, which has struggled to define the engineer, the engineer's "social role," social responsibility, and involvement in public affairs.

The first effort to form an engineering society in America occurred in February 1839 when some forty engineers met in Baltimore.   The constitution did not meet with the approval of the American Civil Engineers and, as a counterproposal, independent regional societies developed.   The first of these was the Boston Society of Civil Engineers founded in 1848, followed by the American Society of Civil Engineers in 1852 which, despite its name, functionally operated as a local society for New York.   Of all these societies, the American Society of Civil Engineers claimed to represent all American engineers not in military service.   It maintained high standards of membership that drew a sharp line between the elite of professional engineers and all others. Various authors have documented the evolution of ASCE and its battles for excellence and preeminence, particularly with the more heterogeneous American Institute of Mining Engineers.[27]

Defining the professional role of engineers and spirit of the engineering societies underlies the continuous struggle to develop standards of professional practice and excellence.   The establishment of a National Academy of Engineering in 1962 under the charter of the National Academy of Sciences, for example, underscored the importance of an advisory position to government agencies and increased the stature of engineering in the U.S.

The American Association of Engineering Societies (AAES) is a more recent attempt to unify engineers. It moved to Washington, DC in 1985 in an attempt to influence public policy. Joining together the Institute of Electrical and Electronics Engineers, the American Society of Civil Engineering, the American Society of Mechanical Engineers, the American Institute of Mining, Metallurgical and Petroleum Engineers, and the American Institute of Chemical Engineers, AAES was envisioned as an umbrella organization and source of information on engineering-related issues. As a result of "turf" wars, the Association ended up as a coordinating mechanism rather than the strong voice of engineering originally desired. The lack of engineering unity has been cited as part of the difficulty experienced in addressing problems such as acid rain, chemical spillovers, hazardous waste, ocean level rise, and other large environmental and social problems.

Applying engineering to societal problems requires an atmosphere conducive to technological development and engineering activities. Three conditions tend to contribute to such an environment:

(1) broad social acceptance of technological advancement (i.e., technology is seen as beneficial);
(2) acceptance by the establishment (i.e., minimizing engineering as a threat to other groups' interests or values); and
(3) a market structure that facilitates the spread of engineering products.[28]

While these factors may not be completely in place, four simultaneous transformations are affecting various professions, but especially engineering. The first is a *conceptual* transformation, that is, new concepts, new theories and new paradigms characterize the field. The second is *methodological*, specifically the presence and implications of increasing computational prowess and multiobjective, multipurpose analysis. The third transformation is *organizational*: the increasing movement from defense to nondefense plants, new management schemes, and administrative procedures. Finally, the fourth major transformation is *substantive,* focusing upon the technological breakthroughs and opportunities now appearing.

In the complicated world that surrounds conversion, engineers, like many other professionals, can be agents of change and need to be adaptable to the changing environment. Historically, engineers have adapted successfully to changing circumstance. Many characterize the current system as highly uncertain, in which change will not only continue but accelerate.[29] Therefore, engineers are well advised to actively prepare for wider roles and for a professional posture that emphasizes variety and challenge, entrepreneurial opportunities and problem-minded rather than answer-giving training and practice.

Standing in the Community

"Public support for science and engineering depends not so much on the discoveries and inventions produced, but on how closely the values of scientists coincide with those of the larger society."[30]   Traditionally, even civil engineers have been reluctant to undertake any major role in public policy development. The special study of ASCE on the "Future of Civil Engineering" concluded that this reluctance stemmed in part from a perceived or potential conflict of interest because such policies are usually directed toward public-sector organizations which are also major employers of civil engineering services. Engineering education has also been criticized for not preparing students to deal with public policy issues. In contrast, attorneys, physicists, economists, and social and political scientists frequently assume policy leadership roles.[31]

Since its founding in 1853, ASCE has tried to bridge the gap between engineers and other professional communities by dealing with a variety of professional, technical, and policy issues.   Similarly, other engineering societies have become involved in policy matters, especially engineering education and research. They have also established Washington offices and hired professionals with federal policy expertise, coordinated policy activities by forming coalitions with other engineering societies and organizations, and supported the development of technical peer review programs at the federal and state level including quality assurance standards for the profession.[32]

Empowerment and Commitment

The image change (Table 1) of the engineering profession from the lone inventor to the manager of complex interdisciplinary projects has implications for the process of conversion and for future constructive participation of the engineering profession in more collective endeavors. For conversion to work we need to ask not only where engineers have been, but where are they going.

Economic conversion will add to the presumed "crisis" in engineering. Engineers will need to move beyond the "crisis response system" to develop a more proactive "risk management orientation" of informed changes and concrete suggestions.

Engineering societies face two major problems: expanding their role and constituency and improving public perceptions of engineering. The wide range of committees of the American Society of Civil Engineering represent one attempt to overcome these problems.   They include professional committees focusing on minority programs, public involvement, recruitment of younger members, employment conditions, professional registration and standards of practice and a variety of technical activity committees (TAC), ranging from cold regions engineering to forensic engineering, including earthquake preparedness, energy conservation and use, materials, highway, urban planning, transportation, environmental assessment, water resources, and geotechnical engineering. More important, ASCE and other engineering societies have been inviting non-ASCE scientists from other disciplines, including economics, political science, geography, and sociology to be members of such committees. As a result there has emerged a direct and indirect influence network, although not a unified engineering voice.

The Conversion Process and the Future

Prospects and Options

There are many ways of describing the last 50 years of momentous changes in terms of political, social, economic, and technological alterations. Perhaps two concepts best summarize the great transformation of the last half of our century, that is, *complexification* and *rapid change* resulting from the great expansion of the role of the government, the rapid increase of data, information and knowledge available, the accelerating rate of technological development, and the globalization and internationalization of the economy and the marketplace. Such factors and their changing impact severely alter the scope and scale of the engineering profession in terms of both the education needed and future professional involvement.

If we examine first the questions inherent in the metaphor of "from swords to ploughshares" then we can examine the extent to which the established structure of engineering can absorb the stresses and strains that are likely to accompany conversion. Most importantly, the educational system will face increasing pressures for specialization and fluctuating demands for engineers in various fields. The curriculum will need revision to integrate technological capabilities with sensitivity to shifting civilian preferences, needs, and demands. The profession of engineering now requires a visionary capability that would produce a systems professional and a practicing individual who would have the foresight, imagination, solid background, and experience to combine data and judgment. A curriculum to prepare such engineers requires the capacity of applying innovation and knowledge about a wide variety of technologies and institutions and the personal capability of leadership in a global community.[33]

Table 1

A Summary of Engineering Role Changes

CURRENT AND PAST	EMERGING
Primary works in manufacturing sector	Continues in manufacturing but now also works in service industries
Male Caucasian	May be a woman, Asian, or foreign-born, with some African- or Hispanic-Americans
Sole earner and head of the family	Member of a dual-income family with higher income and greater financial independence
A technical expert who solves a problem, writes a report	Manages people or a process, must communicate results to others in writing, graphically, and orally
Has a B.S. degree or worked his way into an engineering job	Has a B.S. or M.S. or may have begun with a two-year technician's degree, earned B.S. on the job. Could have entered work force with a Ph.D., especially if foreign-born
Has had one or two refresher courses since college	Participated in at least one retraining program. Takes one or more skill- or knowledge-acquisition course annually. Is studying management skills
Worked for the same employer, a large corporation, for more than 20 years	Worked for several employers, in small and large companies
Is an employee	Is self-employed
Is invisible to the general public—a technical expert in the back room	Well-known and acknowledged as an expert on technological issues and their social impacts; frequently consulted
Has moved up socially by becoming an engineer	May have motives for entering engineering other than social position
Reaches the top executive spot in his corporation with "hands-on" engineering skills as a base	Is studying management and financial skills to gain an advantage in promotion outside of engineering
Is rapidly becoming obsolescent, may retire early, if company offers	Expects rapid job obsolescence and job change; looks to employer and engineering society for career planning and retraining opportunities
Does not or only occasionally uses a computer	Expects to have the latest electronic equipment available on the job, at home, or on the road

Source: J.F. Coates, *Forces Shaping the Future of the Engineer and the Engineering Profession* (Washington, DC: ASME, 1987): pp. 7-8.

The exploitation of new challenges, the capacity for synthesis, the visionary capabilities of both a specialist and a generalist, all require a manpower that is differently trained. Similarly, the professional engineering societies should not only enhance the professional status of their members, but they should advocate visionary action appropriate to practice, education, and research and should play a leadership role in policy formation.

Adapting to an Unpredictable Environment

The engineering profession and efforts to develop competitive high-tech civilian markets have been characterized as "beating swords into...chips."[34] The weapons industry is no longer capable of acting as the most important technology driver. In 1960 U.S. research and development accounted for one-third of all the moneys spent on R&D in Japan and the West versus about one-seventh of R&D today.[35] To retain world leadership in high technology, the Council on Competitiveness believes that the United States must shift from defense-only research and focus more on commercial projects. This implies: (1) encouraging scientists and engineers to identify commercial applications throughout the R&D process; (2) pushing industry to take advantage of the ideas, technology, and talent in federal labs and universities; and (3) breaking down the barriers that prevent weapons makers from buying commercially available products.[36]

In making such a desired transition and fostering strategic alliances, there must be certain unique and effective approaches in the building of a triangular relationship between defense, universities, and private companies. One way is to build satellite research laboratories near campuses so that all three partners can benefit from each other's presence. This is the solution advanced by the Defense Advanced Research Projects Agency (DARPA). They suggest the transfer of technologies such as supercomputers, graphics workstations, artificial intelligence, composite materials, and digital gallium arsenide circuits into the commercial world.

The experience of the Soviet Union demonstrates that economic conversion is not an easy task. The Soviet Union cut its military budget and production after concluding treaties with the United States that radically reduced weapons stockpiles and conventional forces in Eastern Europe and Asia. This conversion is proceeding very slowly and the centrally planned Soviet economy is not any better prepared for efficient large-scale conversion of the military economy than free market economies. An example cited in a recent story in the *Technology Review* of July 1991 illustrates how poorly Soviet conversion is working: the government told the aircraft maker Ilyushin to produce spaghetti machines instead of more passenger planes. Similar difficulties have been reported in England where a report commissioned by Britain's Parliamentary Office of Science and Technology recommends adoption of the "American approach," i.e., an emphasis not on defense spin-offs but on a dual-use technology for both the defense and nondefense sectors.[37]   Governor Hickel of Alaska proposes such a peacetime project under the catchy slogan "Big projects define a civilization.   So why war—why not big projects?" The big peacetime project referred to is the 2000 mile underwater pipeline carrying water from Alaska to thirsty California.

The likely impacts of conversion and the emerging global economic order underscore the larger problems that mark outdated strategies, short-term thinking, technological weaknesses, and government and industry at cross-purposes.[38]   Others who want conversion envision a peaceful economy based on some political, social, economic, and ecological dimensions that emphasize democratic, cooperative decision making, small-scale decentralized production, enhanced local or regional self-reliance, respect of the world's resources and fragile ecosystems, little if any pollution, and heightened awareness of the global consequences of local economic activities.[39]   For these observers, the engineer of the future should be a person who is capable of acting in a sociocultural milieu that emphasizes broader economic principles of sustainable development. The process of conversion must be more widely understood within these terms, but there still remain important omissions and gaps in our knowledge as to how one can make the transition to a new socioeconomic order.

Three issues are top priority: (1) a better estimate of the dependence of the engineering profession on military-oriented activities; (2) a better understanding of the training and curriculum required to educate an adaptable engineer who will become a problem solver rather than an answer-giver; and finally (3) a model of how we can make the transition to a new economic order with minimal disruption to present institutions and with minimum costs to the present work force.

Economic conversion implies a transformation of engineering theory and practice that emphasizes complex systems in macro-engineering, fosters the capacity to monitor and evaluate the social system by building management and decision support systems, and provides continuous efforts for restoring and rehabilitating the system. The suggestion that appears over and over again in many studies on engineering and society and on the future of engineering is that a new overriding requirement in engineering education is to prepare engineers to conduct their professional activities with a greater awareness of their social responsibilities. If engineers are to cope and adapt successfully to likely future conditions, new curriculum requirements for broad engineering education must be introduced with emphasis on non-technical knowledge.   This includes decision-support systems, personal career management, and education.

There still remain a number of unanswered questions that arise from developing a capacity for adapting to an unpredictable future environment. (1) Are styles in engineering associated with socioeconomic, cultural, or ethnic characteristics?   (2) What is the relationship between science and engineering and how does one move from one to the other ? (3) How can we study shifts in career paths and why are such shifts taking place? (4) Is the defense establishment attracting proportionally more or the best of the engineering profession? (5) Are personality characteristics or predispositions significant to various defense and nondefense careers? (6) How do engineers make education, training, reorientation, and other career choices? (7) Does the emphasis on high tech in recent years influence the recruitment, retention, and satisfaction of engineers? (8) To what extent does the function and knowledge of engineers become obsolete and how does one fight obsolescence?

At the heart of such questions and concerns is the hope for a new engineering breed sensitive to a range of issues that go beyond the boundaries of traditional practices and responsibilities of the profession.[40] The next generation will have to be prepared to think transnationally, if not globally, as multinational corporations and far-flung enterprises become the norm. At the same time, future engineers may be held more accountable for the impact of their activities even as they are also asked to contribute to the rehabilitation of a decaying American infrastructure, the containment of nuclear waste, and the correction of failures of previous technologies. They will also have to understand the public and political responses to such problems and, therefore, be capable of joining efforts with other societal constituencies and special interest groups in determining the sagacity of a proposed project. The traditional image of the engineer has been changing not only because of new circumstances but because the new breed of engineers recently graduating has been steeped in the atmosphere of social crisis. They are children of the ecological movement and see themselves as citizens carrying the emerging new conscience of coexistence in a fragile environment.

Sociological models of economic conversion can focus on different levels of society. Such models range from *individual help,* where improving the skills and reorienting individual engineers is emphasized, to *firm orientation,* where particular companies are aided in converting to nondefense activities. In *community's assistance*, larger communities highly dependent on the defense sector are given capital and other incentives to help smooth the transition to nondefense-oriented work and, in *global reordering*, massive social and economic rearrangements follow international agreements for disarmaments and joint ventures.

Economic conversion may hasten the transition from the old practicing and entrepreneurial engineer to the new model of a manager and systems engineer.[41] In practical terms the response of the profession is a renewed recognition on the *logos* in technology and a deeper understanding of the importance of the nontechnical environment. The process of economic conversion is a social transformation that will require a veritable revolution in thinking and practice of engineers and presents a severe challenge to the educational, political, and social institutions that structure our lives.

Notes

[1]Alvin Rudoff and Dorothy Lucken, "The Engineer and His Work: A Sociological Perspective," *Science* 172 (11 June 1971): 1103-1108.

[2]Robert B. Reich, *The Work of Nations: Preparing Ourselves for 21st Century Capitalism* (New York: Alfred A. Knopf, 1991).

[3]Lloyd Jeffry Dumas and Marek Thee (eds.), *Making Peace Possible* (Oxford: Pergamon Press, 1989).

[4]Richard R. Nelson, *The Moon and the Ghetto* (New York: W.W. Norton and Company, 1977).

[5]Francis Fukuyama, "The End of History?" *The National Interest* 16 (Summer 1989): 3-18.

[6]Lloyd Jeffry Dumas, T*he Over-Burdened Economy: Uncovering the Causes of Chronic Unemployment Inflation and National Decline* (Los Angeles: University of California Press, 1986).

[7]Ibid.

[8]Ibid.

[9]Ibid.

[10]Nicole Ball, "International Experience of Conversion," in L.J. Dumas and M. Thee (eds.), *Making Peace Possible* (Oxford: Pergamon Press, 1989), 17-22.

[11]Committee on the Education and Utilization of the Engineer, *Engineering Education and Practice in the United States* (Washington, DC: National Academy Press, 1985).

[12]Panel on Engineering Interactions with Society, *Engineering in Society* (Washington, DC: National Academy Press, 1985).

[13]Theodore Ziolkowski, "The Existential Anxieties of Engineering," *The American Scholar*, Spring 1984: 197-218.

[14]Panel on Engineering.

[15]Bill O'Neill, "Who Wants to be an Engineer," *New Scientist* 126, No.171 (5 May 1990): 42-47.

[16]J. Morton Davis, *Making America Work Again* (New York: Crown Publishers, 1983).

[17]Richard C. Atkinson, "Supply and Demand for Scientists and Engineers: A National Crisis in the Making," *Science* 248 (27 April 1990): 425-432.

[18]John A. Alexander, "The Civil Engineering Shortage: Reality or Myth?" *Education and Continuing Development for the Civil Engineer: Setting the Agenda for the 90s and Beyond*, Proceedings of the National Forum, Las Vegas Nevada, 17-20 April 1990 (New York: American Society of Civil Engineers, 1990): 463-468.

[19]U.S. Congress Office of Technology Assessment, "Preparing for Science and Engineering Careers: Field Level Profiles," Science, Education, and Transportation Program, Staff Paper (21 January 1987).

[20]Ibid.

[21]John Walsh, "Foreign Engineers on the Rise," *Science* 239 (29 January 1988): 455.

[22]Ibid.

[23]J.F. Coates, *Forces Shaping the Future of the Engineer and the Engineering Profession* (Washington, DC: American Society of Mechanical Engineers, 1987).

[24]Ibid.

[25]Edward T. Layton, Jr., *The Revolt of the Engineers: Social Responsibility and the American Engineering Profession* (Cleveland: Case Western Reserve University, 1971).

[26]Henry Elsner, Jr., *The Technocrats: Prophets of Automation* (Syracuse, NY: Syracuse University Press, 1967).

[27]William H. Wisely, *The American Civil Engineer 1952-1974* (New York: American Society of Civil Engineers, 1974).

[28]Panel on Engineering.

[29]Denis L. Johnston, "Engineering Contributions to the Evolution of Management Practice," *IEEE Transactions on Engineering Management* 36, No. 2 (May 1989): 105-113.

[30]Atkinson, 425-432.

[31]Task Committee to Plan Conference on Civil Engineering Research Needs, *Civil Engineering in the 21st Century: A Vision and a Challenge for the Profession* (New York: American Society of Mechanical Engineers, 1988).

[32]Ibid.

[33]Ibid.

[34]E.M., "Beating Swords Into ...Chips?" *Science* 252 (April 1991): 22.

[35]Special Issue, *Fortune* (Spring/Summer 1988): 60.

[36]Ibid.

[37]Jane Bird, "UK Cold War Warriors: Out in the Cold?" *Science* 253 (5 July 1991): 26.

[38]Michael Dertouzos et al., *Made in America: Regaining the Productive Edge* (Cambridge, MA: The MIT Press, 1989).

[39]Dumas, *The Over-Burdened Economy*.

[40]Dorothy S. Zinberg, "The Next Generation of Engineers: A New Breed?" *Aspen Quarterly* (Spring 1990): 50-57.

[41]Robert C. Semans, Jr., and K.F. Hansen, "Engineering Education for the Future," *Technology Review* (February/March 1981): 22-33.

10

Economic Conversion and Global Justice: The Moral Issues

Andrew Jameton

> And what rough beast, its hour come round at last
> Slouches toward Bethlehem to be born? —Yeats

Abstract

This chapter offers some preliminary responses to questions about the relationship of economic conversion to justice and the technical community.

How will the technical community be impacted by economic conversion? Conceptions of justice vary, thus the fairness of conversion for engineers will inevitably be controversial depending on what model of justice one sees conversion to support. Conversion need not necessarily treat engineers justly on any criterion of justice, nor will it necessarily encourage justice globally. However, a conversion process satisfying a complete and complex conception of justice is more likely to enhance justice globally than is a process employing simpler and more abstract conceptions of justice. In order to meet minimal moral criteria, conversion must be just to engineers and others impacted by conversion as measured by at least some criteria of justice.

How is the technical community likely to impact economic conversion? Since ethical viewpoints are little studied, to know how conversion will interact with engineers' sense of fairness is difficult. However, we can speculate that most defense engineers will feel that conversion treats them unjustly and will probably use justice arguments to oppose conversion. They are likely to use these arguments to defend the status quo. They are also likely to use different justice arguments at different stages of conversion and in reaction to different models of conversion. Since none of these positions is conclusive and engineers have conflicting views of justice, sharp controversy within the technical community can be expected.

How can the technical community be encouraged to participate constructively in economic conversion? What are the barriers? (A) Key to obtaining support of the technical community is continuing education, publicity, and projects that (1) engage engineers' idealism, (2) foster international and global perspectives and exchanges among engineers, and (3) help them in a positive way come to terms with their own conflicts over justice, in particular, the contradiction between their individual and corporate ideologies.   (B)   Also, it is extremely important: (1) to treat engineers justly, (2) to account for that treatment in terms that clearly refer to conceptions of justice, and (3) to involve defense-industry engineers in planning for conversion.   Barriers include the costs of education, the potential for conversion to generate unfairness, and the preference of the technical community in general for concrete nonphilosophical issues.

What are the key gaps in our knowledge, key information needed to answer more accurately, more confidently, and more completely the preceding questions?   We need first to learn more about engineers' sense of justice and to discover how interested engineers are in justice issues— especially the ethics of macro-economics and war.   We need to understand the complex interaction of maintaining military forces with economic activity in order to make predictions about the global impact on justice of reduced arms expenditures. Conversion planners also need to develop a global philosophy of justice that is more complete and coherent from a philosophical point of view.   Development of new concepts of justice to meet the present crisis of overpopulation, pollution, and resource limitation is urgently needed.

War and Justice

War and its attendant preparations impose some central dangers facing humanity.   Even a small portion of the world's nuclear capacity could cause so much destruction as to make the concept of war obsolete.   In 1991, the world spent close to $1 trillion on armaments and armies.[1]   The world's military preparedness, engineered in its various forms of destruction, endangers the lives, health, and happiness of millions of people.   When nations are not making war, their preparation for it wastes work and resources that could otherwise be used to improve human welfare.   It makes no sense for our rapidly unifying world to maintain these activities, and yet maintain them we do.

Can the world abate its propensity for war?   To do so requires worldwide revision of deeply held beliefs and practices.   First, we need to conduct the affairs of the world safely and humanely without war, or at least, at a greatly reduced level of armed conflict.   Second, we need an economic vision of how to replace military development by peaceful, sustainable, and worthwhile economic activity.

As defined in the introduction to this book, conversion is defined economically rather than ethically as a shift of resources from military to civilian production. This definition does not tell us what sorts of civilian production should be sought. Envisioning conversion is complex. This complexity arises because of the need to shift from an economy based on growth and expansion to one based on minimal use of nonrenewable resources and directed toward long-term sustainability for the sake of future generations. There is every reason to take seriously the warnings of such groups as the Worldwatch Institute and the Club of Rome: The world has only a few remaining decades of flexibility during which we may still have opportunities to avert a grim and rapid decline in worldwide population and productivity. In the developed nations, to convert military production to ordinary civilian production of consumer goods would simply add more nonrenewable fuel to economies already overblown in their use of natural resources and their need for expansion.

The economic definition of conversion carries an implicit ethical grounding if we perceive the challenge of conversion as a shift of resources from military use to the improvement of human welfare with a minimal depletion of nonrenewable resources and minimal production of pollutants. Not everyone interested in conversion need view it this way. Not everyone is convinced that grim futuristic scenarios of ecological disaster are correct. Many don't care what may happen forty to one hundred years from now, even if they do believe these scenarios, nor are there enough arguments to convince everyone of the interests of future generations. Instead, most of those concerned with conversion are more worried about immediate and concrete economic problems—stagnation, unemployment, poor productivity, declining competitiveness in world markets, declining profits, and increasing poverty. They hope that conversion will reduce taxes and thus improve competitiveness, free capital for new investment, and permit government investment in new programs to stimulate production. And indeed, First World countries, such as Japan and Italy, with proportionately lower military budgets appear to be doing better in manufacturing productivity and growth than countries with proportionately larger military budgets such as the U.S. and the former U.S.S.R.[2]

Deciding to undertake economic conversion depends in part on what we envision as conversion, and in part on whether we think we can make the required changes. But whatever arguments one makes for conversion, to be most attractive, conversion needs to work on as many levels as possible—political, economic, philosophical, and moral. It needs to work both globally and locally. The point of this chapter is to explore how moral consequences of different models of economic conversion can bolster chances for its success. How can ethics provide a means to measure conversion's usefulness to society? Is conversion a justified and ethical enterprise, or are some parts of conversion more justifiable than others?

In addressing these questions, conversion is discussed in relationship to justice. First is a discussion of why justice is emphasized as a moral concept rather than other moral concepts. Three models of justice and three models of conversion, each closely associated with one of the models of justice, are then outlined. It will be evident that I favor the "global" model of justice described herein and its associated notion of conversion over the other two models of "market" and "welfare" justice. The global model is favored, not from any prior adherence to a particular theory of morality, but from worry over the future of humankind during the next forty years.

A second agenda of this chapter is to examine the moral impact of conversion in relationship to "the technical community," especially scientists and engineers working in the defense industry, whose lives, livelihood, and philosophies would be greatly affected by conversion.

What moral conditions must conversion meet to treat the technical community justly during conversion? What is this community likely to think about the justice of these changes? What can be done to facilitate a just transition from military to civilian manufacture for engineers, scientists, and others? What will some of the barriers, from an ethical point of view, to conversion be, and what more do policy makers need to know about the ethics of conversion and the ethical views of the technical community in order to conduct the ethical controversy over conversion intelligently? This chapter attempts to respond to these questions.

Why Emphasize Justice

A wide range of moral concepts is applicable to conversion. However, the limited space here permits either superficial discussion of a wide range of concepts or a more extensive analysis of one concept. An extensive discussion of justice is chosen, because it is a key measure of the moral acceptability of economic systems in our current global situation. Some may disagree with the emphasis on justice, but the definitions and models of justice that follow do not depend upon the reasons for focusing upon justice as a concept.

First, the main concepts that compete with justice in ethical theory—utility and autonomy—may no longer characterize ethical goals appropriate to our present global situation. Justice should be favored over utility at this juncture because, in its usual application, utilitarianism favors social policies that aim to increase individual happiness. Happiness is usually seen as increasing when the material welfare of the population increases, as measured by such monetary measures as GNP or average wages in relationship to prices. However, in a sustainable global economy, this close association of happiness, material welfare, and dollars cannot be maintained. The world is reaching a new point of global scarcity that requires a shift in older concepts of scarcity. In the past, an important option for coping with scarcity was to increase our use of resources. But as the human population approaches the carrying capacity of the earth, humans will reach some level at which an average increase in material welfare has to stop. In other words, the consequence of a sustainable economy is that material welfare is represented by a flat line at the point of maximum development. Under these circumstances, increases in average happiness need be achieved without increasing the use of materials. This would of course have consequences for monetary measures in relationship to happiness. The doctrine of progress through increasing material wealth as measured by increased average incomes and increased GNP must come to an end in the coming decades. As the primacy of the material welfare model shifts, aggregative monetary measures will also need to change. Measures that show the distribution of existing goods and equity of access will become increasingly important.

The concept of autonomy should receive similar treatment. The fundamental U.S. economic conceptions are based on the autonomy of individuals in improving their lot through labor and use of resources. In Locke this is quintessentially the conversion of natural resources for private use. This process can no longer be viewed in isolation, and therefore, autonomy can no longer underlie our notion of economic freedom. Instead, individual enterprises will need to be bound strongly by national and international policies on resource use. Such policies will all need to be conceived as fair.

Second, reducing the incidence of poverty is essential to both justice and human welfare in general. In the Third World, poverty supports population expansion, destruction of natural resources, and immense human suffering. In the First World, poverty creates bitterness over lack of opportunity and participation while fostering many of the social problems of modernity—violence, substance abuse, family disorder, and poor health. Poverty also makes the achievement of happiness difficult since it increases suffering and decreases the capacity of individuals to engage in responsible and meaningful activity.[3]

Third, although controversial, the extent of private wealth must also be reduced. This is partly a mathematical consequence of the elimination of poverty in a no-growth world economy. Poverty cannot be eliminated and overall average income stay limited unless the numbers of the wealthy are also reduced. This is especially the case since the wealthy are so much richer than the poor. Incomes of the richest fifth of the population now measure approximately fifty times the income of the poorest fifth.[4] Moreover, it is no longer merely unfair that some are much richer than others. Now, the rich have become dangerous to the rest of the world. Like the world's very poor, the rich destroy resources. They drive cars; fly in planes; support energy-intensive manufacture; heat large houses; make trash; eat meat; and generally enjoy sins which, more than exciting envy, destroy resources needed for the elimination of poverty. A few rich people will find it more difficult to develop a sustainable economy than many people materially less well off.

Thus, in being concerned with conversion, policy makers must be concerned not simply with increasing civilian resources. Instead, planners need to combine increased civilian job opportunities with decreased use of nonrenewable resources. Planners need to find ways to break the link between happiness and wealth.

Definitions of Justice

The term "justice" identifies a broad and fundamental ethics concept. It also can be used to assess the relationship of individuals to each other, as in marriage or between employer and employee. It can also apply more broadly to an individual's situation in relationship to the world at large or to a class of persons. For example, an adolescent with cancer who is confined to the hospital might feel that he or she was unfairly picked by God or nature to suffer while friends go to the prom and on to college.

Similarly, the term applies to relations among institutions, as between IBM and its competitors, or among nations, or most grandly, as a general assessment of the world; for instance, "the United States is founded on an ideal of justice for all," "the world is not fair," etc.

Justice is an old conception and has suffered many different formulations over the centuries. One of the oldest versions is that of individual deservingness:[5] Justice as deserved applies when someone receives appropriate punishment for wrongdoing, or in the obligation to repay debts, and compensate according to merit or seniority. Reciprocity is also an ancient and key conception of justice often recognized by anthropologists in their accounts of relations within families and small communities. This conception emphasizes equal exchange between people, as, for example, when one person imposes an unclaimable but nevertheless legitimate obligation on another by giving that person a gift.

"Distributive justice" moderates the fair sharing of scarce resources between individuals on broad macro-economic questions about investment of basic resources. Thus, "you cut the cake, and I choose the first piece," may operate as a principle of justice between two children, while in medicine, rationing through a complex national organ-retrieval system may assure fair access for those most in need of scarce livers and hearts.

Most modern conceptions of justice emphasize egalitarianism. The principles of equal treatment and equal opportunity dominate much of our political rhetoric. The principle of equality maintains the conception (1) that each person initially counts equally in his or her entitlement to the basic goods of society and (2) that principles and activities modifying this entitlement need to be formulated justly and be applied to all equally.[6] This modern notion of justice contrasts with more ancient conceptions, which attempt to harmonize relations in a community or between individuals without reference to equality.

Justice can be distinguished from notions of general happiness, beneficence, and human welfare as ethical principles in that human welfare and happiness are *aggregative* concepts, where the decision maker chooses the action that maximizes human welfare as a whole. Justice is a *distributive* concept. In contrast with the aggregative principles, justice usually prefers small gains to many over large gains to a few. Moreover, to assess the justice of an action, we need a *comparison* between groups of people, so that a judgment about distribution can be made.

The impact of technology on justice must be looked at in relation to who controls its development. New technological developments have sometimes increased the potential to enjoy life and equalized the power and status of individuals: books, electric lighting, sanitation, roads, and public transit all contribute to the common good in roughly equal ways. But, this need not be so. A luxury-designed airplane may be available only to the wealthy while its noise reduces the quality of life of inhabitants of the neighborhood of the airport; technologically sophisticated and therefore expensive armaments favor wealthy nations over poorer ones. Thus, we can formulate the issue of justice in conversion by asking: (1) whether military technology is being put to purposes that increase or decrease power sharing, and (2) whether the proposed benefits of conversion will increase power sharing. For instance, suppose that a fictional weapons manufacturer is producing a type of antipersonnel weapon particularly useful in terrorizing peasants. Suppose that as a result of conversion, two effects are obtained: first, the oppressors of the peasants turn to less efficient weapons and as a result the peasants win their freedom (i.e., justice is increased); and second, the conglomerate turns its assembly line to making monitoring devices used widely by large companies to observe and control their employees (i.e., justice is decreased).[7] Much more social justice would have been achieved if the fictitious corporation had chosen to manufacture equipment for public transit and housing for the poor. Conceivably some conversion projects would produce less justice than current military projects, for instance, by creating a profound economic depression in some geographical areas.

Before we can assess the prospects and alternatives for conversion we need more specific models and criteria of justice by which we might judge the justice impact of conversion plans. Once planners get more specific criteria for justice, they can then think about what to look for in various models of conversion planning. Since conflicting social and political groups hold different competing principles of justice, planners will doubtless find that almost any model of conversion satisfies some models of justice and not others with the result that almost any model of conversion will prove to be ethically and politically controversial, even if planners can agree on the facts and predicted outcomes of various proposals.

Models of Justice

Free-Market Justice

This model puts the onus of achievement of justice on a fair exchange between people. The thesis is that, if all exchanges between people meet certain criteria, we will have a just society as well as one in which human welfare is maximized. The theory does not expect that this process will result in an equal distribution of goods, because the distributions of goods, equal or not, are unstable and that the fairness of individual transactions will outweigh any differences in distribution. A fair transaction is one that is undertaken freely by both parties, and where both have good information about what they are buying and selling. To a degree, this conception of justice echoes the ancient conception of reciprocity but with a difference: Reciprocity embodies a conception of equal exchange whereas free marketing does not require exchange to equal advantage. The strength of the theory is that it requires no political process by which to agree on values or priorities—social priorities result from individual value commitments realized in transactions. In theory, no one is coerced and thus liberty is maximized and universally distributed even if ultimate overall distributions are unequal.

Social Welfare Justice

The free-market model of justice has been widely criticized. A "social welfare" model of justice argues, for instance, that buyers and sellers are usually not acting freely: the poor are especially coerced by their needs and lack of resources, and in general parties with different resources bargaining with each other are not equally empowered. Moreover no one can be completely knowledgeable about transactions and so those without knowledge need protection from those who would exploit their ignorance. The welfare model thus proposes that, although markets work to distribute a variety of goods, markets must be supplemented to permit allocation of basic goods for those who are not as well off in the market system. Thus keys to the social welfare model are basic income supports, universal health care, and good public education programs. The welfare model captures the notion of justice that no one should be unduly deprived and emphasizes bettering the condition of the worst off in society as a major priority. It also favors equality by ensuring that no one is too greatly disempowered. However, it does not emphasize equality so strongly that it would eliminate aggregations of private wealth.

Global Justice

A "global" model of justice attempts to consider justice in the context of our present global situation. This model has two advantages for us here. First, it involves considerations appropriate to our present difficult global situation in which we face potentially overwhelming climate, resource, economic, and population problems. It thus introduces a global perspective much needed in discussions of justice.

Second, the model follows a recent trend in discussions of justice in health care. It recognizes that specific principles of justice may differ from one area of concern to another. And it provides a list of several characteristics of justice in operation with reference to the particular area of concern.[8] By pulling a number of considerations together, it introduces a handy list of criteria for use in assessing models for conversion.

The model also attempts to express something of the Platonic idea of justice articulated in the *Republic*. Platonic justice seeks to achieve harmony in a community. Plato held that this was accomplished by each person, or each social class, doing his or her own job. Society's members were expected to find meaning in their work through a sense of making a contribution to the community as a whole. Plato's model thus supports engineering ideals that see the profession as meaningful through its overall social contribution. The global model integrates meaningfulness in work with a conception of global fairness. The model employs five criteria to cover various facets of our global situation.

Intergenerational Equality

If we grant that it is unfair to destroy the life opportunities of future generations, a fair social plan must support a sustainable economy. Alan Durning relates this principle to the Golden Rule:

> Each generation should meet its needs without jeopardizing
> the prospects of future generations to meet their own
> needs.[9]

Thus nonrenewable basic goods should be regarded as scarce, and the consideration of sustainable development must take first priority in assessing the use of such resources.

Meaningful Work

Not only is it important to each individual to have meaningful work, but it also provides a way that increases happiness without necessarily increasing the use of resources. Employment is also a key element in implementing justice, since work provides the opportunity both for people to make a contribution to others and at the same time to receive reciprocal rewards.

Limits

Equality is a central element of justice. No one should be unduly deprived, kept in a desperate state, or put unduly under the power of others. Although it is widely recognized that placing a floor on income is an implication of equality, it is not as widely recognized that a ceiling on income is also implied by equality. Permitting very large incomes for a few individuals and families is inconsistent with equality, because first, sustainable growth requires limited resource use and the large resource costs of the wealthy hinder the ability of societies to set limits. Second, enormous differences in lifestyle create inequality, envy, and conflict even among those who have enough. Third, the possession of great private wealth permits some groups to have power over others. Thus, equality must involve limits both above and below.

Human Rights

Basic rights and reciprocity in individual relations must be maintained.

Liberty

Within the preceding limits, policies should maintain maximum liberty to define individually and communally meaningful activity and work. In contrast to some theories of justice, individual liberty to use productive resources should be subordinated to major global concerns.

Questions and Criteria of Justice

This is not the place to argue the virtues of these three conceptions—market, welfare, and global justice. It is useful, however, to note that these and similar conceptions have been crucial to the political economic debates of the last centuries. Moreover, these models suggest various criteria, not necessarily consistent criteria, but certainly pertinent criteria that policy makers might use to assess the justice of various conversion proposals.

Some key questions suggested by these models are:

What is the likely impact of conversion on the liberty of those involved? Does conversion tend to increase the overall level of liberty within a group or between groups being compared? Would people in the world have more liberty? Would engineers have more liberty as a result of these changes? Would some group suffer hindrances in their liberty, such as those who truly need and desire to make armaments? Is there a group that would be left defenseless or impoverished without access to an arms industry? Would a model of conversion tend to empower some people a great deal, but give others little power or disempower them?

Does conversion tend to increase the welfare of those who are worst off? For instance, would it free resources for those who need income support or health care? Or would it create a new class of impoverished, unemployed engineers and thus swell the ranks of those who are the worst off?

Is conversion likely to move toward a more sustainable economy, or does it replace military production with manufacture of energy-costly, discretionary consumer goods? Are the engineering projects fostered by conversion meaningful and worthwhile in their contribution to the community, or will engineers find themselves removed from meaningful work (defense work is meaningful to many engineers) only to undertake trivial design and production projects?

Will conversion tend to moderate income inequality? Will conversion be basically fair in respecting, insofar as possible, existing agreements and contracts? Will conversion foster, rather than restrict, opportunities for the technical community within the limits set by sustainable use of resources?

Why should planners focus our discussion on "the technical community?" What claims of justice might engineers have in contrast to other groups with relationship to conversion? Would anything in particular be owed to them or expected from them as a consequence of conversion?

While many other questions about the consequences of conversion exist, these reflect the concerns of some major parties in the modern debate over justice.

The Scope of Justice

Perceptions of "what is just" are greatly influenced by the scope of the comparisons. Inattention to scope has substantial consequences for how observers see the problems of justice and injustice. An example is that of nurses relative to physicians. Nurses often compare their situation with that of physicians. In comparison to physicians, nurses are disempowered, ignored, unrecognized, and sometimes abused. Nurses perceive that they are unjustly disadvantaged in relationship to physicians. But, nurses rarely compare their situation with that of nurses' aides or environmental services employees, on whom their work often has significant impact. Nurses do not say, "I am unjustly advantaged in relationship to aides," although aides may feel differently about the issue.

Planners thus need to be attentive to the comparison groups they choose in discussing the fairness of the economic impact of conversion on engineers. At the extreme, as long as there are people living in undeserved wretchedness on earth, anyone who is well off is unjustly advantaged by comparison (as long as we include the wretched in the scope of comparison). Moreover, as long as there are people extremely wealthy and influential in the world as the result of chance, theft, inheritance, and other measures unrelated to deservingness, most people are unjustly poorly off (if we include them in our scope of comparison).

The Focus on Defense-Industry Engineers

In assessing the justice of conversion, engineers will likely limit their scope to comparing their state to those who earn more than they do rather than less than they do. Certainly, engineers will compare their new levels of income with their prior levels of income. Thus, any program that results in less pay for the technical community will be charged with injustice, even if the community is highly paid in relationship to many in other sectors of the economy.

Indeed, as largely well-paid white males, defense engineers represent a relatively advantaged group in relationship to many groups in the U.S. Thus, if the question of justice were simply one of comparative welfare, there would be little reason to be concerned for engineers more than any other group. Some might even suggest that there would be less reason to do so since other groups have been first in line for equity-based judgments for generations before engineers.

Engineers may also feel that they have some special claim based on the history of a long-standing relationship between the public and the defense industry. They may argue, "I have given you many years of devoted service; it is unjust of you simply to cast me off now." But, it could be argued in reply that the salaries and rewards (together with accumulated pension plans) paid out during the years are the just compensation for devoted service, so that no further rewards are owed. Since defense engineers will be affected by conversion, it is important they be treated justly. However, they have no special claims to justice different from others in the defense sector—secretaries, administrators, janitors, truck drivers, etc.

Some argue that scientists and engineers represent an important source of creative energy on which successful conversion depends. Special privileges should then be accorded them. Current information suggests, however, that their contributions are not clear. Specifically, the kinds of engineering needed for a sustainable economy are very different from that employed in the defense industry. For instance, Sivard[10] points out that

> The world's armed forces are the single largest polluter on
> earth; in the U.S. they produce more toxins annually than
> the top five chemical companies combined.

Defense engineers may not have the right track record to re-engineer America toward a nonpolluting, resource-efficient world. Although engineers are key to conversion, successful conversion may require newly trained engineers.

Global Comparisons

When thinking about justice, should planners be thinking globally, or just nationally, locally, or, most thoroughly, all of the above?   Peter Singer, in a famous article "Famine, Affluence, and Morality," published in 1972, enlarged the scope of comparison for justice to the world and included the many starving people.[11]   He argued that as long as many people in the world were starving, no one could justify spending more than was needed for basic care and maintenance.   People with resources should reserve their surplus income for ameliorating the condition of the starving.   Although Singer emphasizes utilitarian arguments, not justice arguments, the justice argument is strong.   Specifically, it is unfair for us to be spending money on dining at fine restaurants, burning gasoline on vacation trips, etc., when we could be expending these resources to correct this global imbalance in suffering.   His is a powerful and compelling argument, all the more powerful for the unwillingness of most of us to give it much heed.

The state of the world is well known[12]:

> One out of every eight people worldwide lives on an income of less than $300 per year, in a state of destitution so total that it constitutes silent genocide. Over one billion people, one-quarter of the world's population, are seriously ill or malnourished. In regions of Southeast Asia, nearly 40 percent of the population is afflicted with malaria, measles, diarrhea and respiratory disease, as well as hunger. In sub-Saharan Africa, where the situation is even more dismal, 10 million children die every year of causes which are easily and inexpensively preventable.

Nor can anyone imagine that these tens of millions in any way deserve their fate.   Many practical reasons prevent us from a rescue course (How does one help?   What really works?   Is my help wanted?) but more crucial here is the issue of the scope of the comparison.

In the present global situation, local policies have a potentially significant impact on everyone, including future generations.   Thus, burning fuel in the U.S. fosters global warming, which effects everyone in the world. Burning rain forests in Brazil similarly effects everyone.   This does not mean that the interests of engineers should be sacrificed by conversion for the welfare as a whole, but it does mean that global welfare should be considered in thinking of the burdens and rewards that everyone faces, and engineers should be included in these considerations.

Practically, one should perhaps focus on those areas of justice for which there exist political and economic groups able to use and articulate a concept of justice in decisions affecting relevant groups. But existing political boundaries beg the question of what groups and institutions should be considered in making resource allocations. For example, residency across a state boundary should not restrict someone who needs an organ transplant from access to that organ. And any serious tradition of respect for human life should support policies that show concern for life worldwide and should not be restricted to U.S. lives.

No one knows the potential political influence of a moral argument until it has been asserted widely and forcefully. Thus, although planners may want to confine discussion to what can be accomplished politically inside the U.S, or to expand their scope of concern to what might be done by the U.N., the GATT, the European economic community, etc., it remains possible for citizens to form groups—political parties, issue-based movements, etc.—that in a few years may have a global impact because of their ability to articulate global principles of justice. In short, planners simply need to make choices about the scope of their concern for justice.

Models of Conversion

Any model of conversion is likely to make the world more just. Even a sketchy glance at the extent to which the world's resources are invested in military activity, while important social needs go unmet, supports conversion. Additionally, new arms technology tends to foster further arms development and a consequent overall increase in the level of armaments worldwide. The price of arms favors the wealthy nations over the poor. As major polluters of the environment, the military and defense industries fail to meet the intergenerational sustainability criterion of a global sense of justice. Conversion must nevertheless meet a few criteria in order to be just on a global scope. Such a conversion must not be economically disastrous or lead to imbalances of power such that some populations are more disempowered in relationship to others than they now are.

Any assessment of the justice impact of conversion will depend on what model of conversion is implemented. At least three models are possible.

The Free-Market Model

With a free-market model, the primary aim is to reduce military production. U.S. citizens will save the taxes associated with the military. U.S. policy would not, however, undertake any specific alternative projects, such as rebuilding the nation's infrastructure or technologies for environmental protection. Instead, policies would encourage engineers and investors simply to make good use of their skills and capital. Local and national governments might support agencies that aid investment in small businesses for engineers, retraining for other employment, and market analysis for investors. Planners would strive to maximize the opportunities of those affected by reconversion and avoid value judgments about what the alternatives should be. In this model, if some workers fail to make the transition despite temporary assistance, this is regrettable but not unfair.

The Compensation Model

Under the compensation model, policies would reflect concern over the consequences of the conversion on those directly impacted by it. Conversion policy would be particularly concerned with the economic impact on companies, engineers, and their surrounding communities. Policy makers would strive to ensure that these companies and communities survive. Local governments would struggle to find opportunities for existing aggregates of capital, projects that employ the existing talents of engineers, and government support for those who can no longer make a contribution. The scope of concern would thus be for the most part on those impacted by conversion and on compensating them for their sudden and undeserved change in status.

The "Global Planning" Model

Under the global planning model, policy would regard conversion as an important ethical opportunity for the world at large to deal with its problems. Conversion could free tremendous amounts of capital for new large-scale projects. These newly released resources could generate energy to address major unsolved problems of the world. This model has several advantages over the others: Large civic projects would likely require the work of engineers, although many of them would need retraining. Ideas for projects are many: public transit, sanitation and waste projects, wilderness creation and maintenance, computerized medical diagnosis, ozone restoration, atmospheric carbon reduction, universal telephone service, and energy efficiency. These projects attract the idealism of engineers to meaningful work in which their technological imaginations would be used to serve the ends of global welfare and justice.

Model Comparisons on Justice Criteria

The free-market model meets minimal conceptions of justice. It maximizes liberty by minimizing the direction of engineers to new projects and by supporting the abilities of engineers to find new work. It thus meets a market model of justice and respects engineers. If technological alternatives are developed readily, then no one will be greatly harmed. Even if engineers fail, the model is fair because it operates by the neutral principles of the market and not because of any unfair treatment to any party.

As measured by additional criteria of justice, however, this model falls short. If communities do not stimulate economic alternatives, many engineers, other individuals, and investors will become impoverished. The wealth associated with the defense manufacturers will disappear because their work is no longer valued and opportunities do not exist for their use. One could still make the case that justice is done overall because the world's security is increased and people around the world are thereby benefited. The sacrifices of the few affected by conversion would thus be justified and made meaningful by other gains. It would be difficult, however, to explain this to the engineers affected.

The model has some additional justice-related problems. Engineers in general could choose alternative work that is harmful to sustainability and global equality, as they might by turning their work to increasing the power of the elite and thus reduce the overall gain in justice. Additionally, if the U.S. removes tax money from the defense industry and places it someplace else, then the U.S. will still not have solved its debt difficulties nor perhaps solved marketability problems in the global economy. For example, even though reformers often make comparisons between health and defense expenditures, U.S. health expenditures are so high and problems of access to health care are so great that putting more money into the existing health care structure would not improve the situation.

Overall, this model offers a useful insight: Even the most limited form of conversion (the free-market model) would likely meet the criteria of a market model of justice in regard to its impact on the technical community. It would also likely have a positive global impact of some significance.

The compensation model is less libertarian than the market model of justice. It does not satisfy some of the justice criteria that were satisfied by the free-market model but it can satisfy other considerations of justice not satisfied by the free-market model. In its social welfare orientation, it prevents much potential harm to engineers and investors in the defense industry. These groups would be protected by the assurance of alternative forms of labor. Many such projects could also support aims of global justice by improving welfare in the U.S. and abroad. Although charges of "make-work" as welfare and of corporate welfare might be made by those paying taxes, services would probably compete favorably with other kinds of work. However, these projects would not necessarily contribute to global welfare.

Communities might well choose projects mainly for their own benefit rather than their global contribution.

The global planning model likewise does not satisfy the minimal market conception of justice. It includes substantial value commitments that override the justice of markets; neither would it necessarily assure the welfare of engineers and defense investors. It might expand markets and it might result in very good opportunities for defense engineers. And it meets global criteria of justice by orienting discussion of conversion options to sustainability and global justice. By involving engineers ideologically, it has more potential for some engineers to feel that they are being treated justly by the changes than do other models.

Engineers' Views of Justice

Engineering ethics in the past have not emphasized justice. Engineering codes of ethics do not usually refer to justice; instead, they tend to refer to the goal of improving human welfare at large, i.e., to aggregating the good, rather than distributing the good. However, the codes do not prohibit considerations of justice. In contrast, by their overall support for virtuous professional conduct, they foster virtue in all its forms including justice. However, by emphasizing the language of utility over the language of justice, the codes permit an engineer to function ethically by benefiting a small portion of the population, even if that work may benefit those special few to an unjust degree in relationship to others.

In fact, it is difficult for engineers, in their position, to address questions of justice in the use of their ideas. For example, if an engineer designs a bridge or dam, the engineer can feel confident that the bridge or dam will be used to some good for someone, especially if he or she designs it safely and sturdily. However, a toll could be imposed on the bridge, unfairly restricting the access of some of the population on one side, or the electricity from the dam may be used to power luxury housing while others are left homeless. Working engineers are in a poor position to control, or even to participate in decisions about, these aspects of a project's use.

Moreover, engineers are like most people. They have a poor sense of justice when their own interest is involved. Like nurses, they no doubt look up to those more wealthy and powerful than they, rather than concern themselves greatly with the fates of those of lower status. Research also suggests that engineers are conservative politically. Their ideas of justice are largely those associated with a free-market approach. Engineers seem to hold these views in spite of the fact that their primary work experience is within large organizations where work is tightly coordinated and provides for minimal liberty.

Defense engineers would thus likely oppose the conversion process; indeed, many are likely to regard conversion as an unjust process. Engineers working in defense have probably come to terms for the most part with their ethical questions about their work and doubtless feel justified in their present situations. Any major change not coming from their own efforts, innovation, or promotion would likely be resented, even though no theory of justice holds that anyone must be continued in their present jobs.

Engineers also doubtless believe that they are where they are through their own efforts and that intelligence and skill determine advancement in the world. Thus, any major change that greatly reorganizes their work and that was imposed from the outside would appear unacceptable.

CEOs, investors, and engineers who have become managers would doubtless hold similar views. As company officers, they are concerned about what happens to their corporation's capital. They expect that protection of it should be one of the duties of society. They perceive that they are entitled to what they make, because what one wins in a free market is always justified. Continuing to make income on capital is important and will continue because of the social good it serves. Social planning and policy are viewed as unjust because they force extrinsic values on the market and make it inefficient. It is better to meet human needs through free enterprise than through taxation (the defense industry excepted).

Thus these groups would likely be ideologically opposed to any model that involves more than the free-market conception of justice. At the same time, they would also be concerned about that model because it gives them the least support. Winners, not losers, in the free market prefer a philosophy of market justice. Thus, we should expect bitter conflict within the defense industry over various models of conversion. Free-marketeers within the industry would be faced with extremely stressful contradictions. Perhaps the only expectable consensus would be general dissatisfaction with the compensation model, which fails to satisfy the free marketeers without any substantial ideological payoff or new markets to replace the old work. Many who might be convinced of the value of maintaining a strong defense together with some of the global objectives may be concerned about displacing one with the other.

In conclusion, even though the technical community is likely to lead with a version of market justice to resist change, and then to move reluctantly to a welfare model if they lose, policy makers can offer the global model of justice as a promise of something better than the other two. In the early part of the century, U.S. engineers took social leadership in envisioning the contributions of technology to modernization of society and raising the standard of living. It may be possible to reawaken that sense of technological idealism.[13]

The Power of Justice

It is important that those concerned with conversion be concerned with issues of justice. First, justice is one of the concepts of ethics with the most powerful appeal. Because justice is a distributive concept, it tends to protect everyone; when the technical community starts to feel vulnerable to conversion, it will place much stake in arguments from justice. Second, any sustainable political community must be accepted by its population as being just. Third, even if people are personally disadvantaged by a plan, they are more likely to accept it in the end if they feel it is just. Fourth, justice suggests that empowerment of engineers in relationship to conversion is important. If defense engineers are widely consulted on conversion plans and if they feel that their views have some impact on the plan, they will be more likely to feel that the plan is just than if they are not consulted. Thus, if conversion planning can be democratized, those affected will be empowered rather than unjustly deprived of power.

Notes

[1]Ruth Leger Sivard, *World Military and Social Expenditures* (Washington, DC: World Priorities, Inc., 1991), p. 7.

[2] Ibid., 27.

[3]Poverty hinders happiness, but above the poverty level, happiness is poorly correlated with material welfare and income. Thus, elimination of poverty is a key element in a strategy for maximizing happiness while minimizing resource use. For a general discussion of resource use and happiness see Alan During, "Asking How Much is Enough," in Lester R. Brown (ed), *State of the World 1991* (New York: W.W. Norton and Company, 1991), pp. 153-169.

[4]Sivard, 5.

[5]See the first part of Plato's *Republic* for Socrates' critique of this conception.

[6]This is from Rawls, but I substitute "basic goods" for "liberty."

[7]The reader who does not like these examples may readily generate other examples increasing and/or decreasing the level of justice in the world.

[8]Reinhard Preister, "Taking Values Seriously: A Value Framework for the U.S. Health Care System" (Minneapolis: Center for Biomedical Ethics, University of Minnesota, January 1992); Paul Reitmer, "Health Care Reform, Justice, and the American Way," Center for Health Services Research Quarterly Report 2 (January/April 1992): 1-2.

[9]Durning, 165.

[10]Sivard, 5.

[11]Peter Singer, "Famine, Affluence, and Mortality," *Philosophy and Public Affairs* 1, No. 3: 229-243; 1971.

[12]Bernard Lown, Eugene Chazov, Forward to Sivard, 3.

[13]Edwin T. Layton, Jr., *The Revolt of the Engineers: Social Responsibility and the American Engineering Profession* (Cleveland: Case Western Reserve University, 1971).

11
Economic Conversion and Engineers: What Do We Know?

Patricia L. MacCorquodale, Martha W. Gilliland, and Jeffrey P. Kash

Abstract

This chapter summarizes the likely impacts of economic conversion on engineers and, in reverse, the impact that engineers are likely to have on conversion. Together, these impacts suggest factors that could increase both the participation and influence of engineers in bringing about a smooth transition.

Impact of Economic Conversion on Engineers

Economic conversion will result in at least four types of impacts on engineers: impacts on the job market, income levels, transferability of talents, and job satisfaction. This section first summarizes the four categories of impacts and then discusses factors that could affect the severity of the impacts in a major way.

Impact Types

Job Market

In the short term, reductions in military spending will result in fewer jobs for defense engineers, unless military cuts focus on personnel and weapons spending is maintained at current levels. However, not all types of engineers will be affected equally. Engineers most severely impacted will be older, highly specialized engineers in the defense sector. The engineering specialties most concentrated in defense-related activities are aeronautical/astronautical, electrical/electronics, mechanical, metallurgical/material, nuclear, and industrial engineers. The group that might be most highly affected is aerospace engineers, of whom nearly half are employed in the defense sector. Unfortunately, without new jobs in the civilian sector created by economic conversion, cuts in the defense budget may create a situation in which more defense engineers are seeking fewer defense positions. This situation will be particularly acute for aerospace engineers, older engineers, engineers doing highly specialized technical work, and engineers whose work requires a security clearance.

Income Levels

Clearly, when fewer jobs are available for the same number of qualified people, salaries and benefits decline. This is a likely scenario for defense engineers, especially in the short term. An extended period of adjustment (15-20 years) may accompany the decline of salaries for engineers. Currently, salaries for defense engineers may be artificially high because cost is not a primary consideration in the defense sector. If defense cutbacks result in large numbers of unemployed engineers, the competition for engineering jobs in consumer production may increase. This situation could result in lower salaries for civilian engineering jobs. The degree to which a drop in salaries will affect engineers who are not in defense depends on how rapidly defense cuts occur, the general health of the economy, the number of new engineers, and the extent to which economic conversion is planned.

Transferability of Skills

How much difficulty engineers will experience moving from military to civilian work is a fundamental question in planning economic conversion. The principal focus for military work is the development of products that will perform under the unpredictable and generally extreme conditions of combat. Therefore, performance rather than cost is the primary consideration. In contrast, civilian work is aimed at developing affordable products. Retraining defense engineers to consider product cost in addition to performance may be difficult.

The transferability problem can be a major hurdle to individual defense engineers looking for work in the civilian sector. The expertise and specializations that are valued in high-technology military design work are not easily applied to low-technology and cost-efficient consumer products. Other issues, such as the security clearance status of defense engineers, may exacerbate the problem if "gag" rules prohibit engineers from applying their skills in the consumer sector. High security clearance delays engineers' opportunities to share the technical knowledge acquired in defense-related work. This may make defense engineers less marketable for consumer-oriented jobs.

The problem of transferability is especially acute for engineers in the national laboratories, where a national security orientation defines a substantial part of the responsibility of the engineering work force. Conversion could require that the labs redirect their focus to civilian R&D. It may also be necessary for labs to attract private investment and to design consumer-oriented products for commercial sale. Changes in the direction for research in the labs and their large price tag raise questions about whether it is desirable or appropriate to keep all of these government-sponsored labs.

If engineering R&D at the national labs is redirected successfully from defense to civilian work, that success may provide an injection of technology into the civilian sector. Conversely, unsuccessful conversion of the national labs will deprive private industry of an important and sophisticated source of R&D. Additionally, the jobs of lab engineers, and quite possibly the labs themselves, will be in jeopardy.

Job Satisfaction

A shift from work in the defense sector to the civilian sector could affect the job satisfaction of engineers. The shift could require adjustments in the value systems of some engineers. Specifically, maintaining national security through defense-related work has been a source of pride for many engineers. A greater appreciation of the importance of economic competitiveness and economic strength as a prerequisite for national security is needed if engineers are to fully understand the importance of the conversion process. Similarly, alternative national goals, such as environmental protection, can be enhanced by advanced engineering technology. Thus, job satisfaction could be enhanced if engineers embrace new values and new endeavors. However, these new endeavors may not provide the high pay that has characterized defense work.

Job satisfaction for engineers is also related to job characteristics, e.g., pay scale, seniority, technical challenge, and job security. Additionally, engineers' opinions about whether economic conversion is bad or good may affect how satisfying they find new jobs provided by economic conversion. The changes in existing job characteristics initiated by the conversion process may be interpreted as unjust by some engineers. If the loss of jobs in the defense industry leads to declines in pay, seniority, technical challenge, and job security (the scenario of rapid and unplanned conversion) engineers are more likely to feel that the impacts of the conversion process are unfair. On the other hand, if planning creates new opportunities in the civilian sector that offer equal or better job characteristics, engineers are more likely to support the conversion to civilian production as a just process. The moral or ethical judgments made by engineers, based on the impact of the conversion process on their lives, will largely determine their attitudes about the conversion process.

Many surveys indicate that engineers experience relatively lower job satisfaction than other professionals, especially as they become more experienced and older. Engineer managers generally experience higher job satisfaction than engineers with similar backgrounds who are engaged in technical work. Because highly technical defense work requires the cooperation of large companies and large numbers of engineers, engineers who work in teams often have less autonomy. In many instances, the development of consumer products requires smaller numbers of engineers due to the less technical nature of many consumer goods. Thus, engineers are likely to experience more autonomy in consumer industries than in the defense industry, which may enhance the attractiveness of economic conversion for them.

The impact of economic conversion on job satisfaction depends primarily on the number and types of new engineering jobs that become available. Most engineers are known to prefer high-technology work and such jobs may be available in the surveillance and medical equipment industries, engineering software development, and joint military-commercial endeavors. Other specialties that could be expected to provide enhanced job satisfaction based on their social value include environmental protection, infrastructure development, and the engineering of products to assist special groups such as the disabled or elderly.

Severity of Impact

Planning and Decentralization

The less planning that is done for economic conversion, the greater the job loss and job uncertainty for engineers who are working in defense jobs. Both good planning and decentralized planning can mitigate the impact. Decentralized planning involves the development of conversion plans for specific military installations and weapons production facilities at the local level. Decentralized planning provides the best starting point for engaging the skills of engineers. By utilizing the skills of engineers who are familiar with a facility's equipment and operations, decentralized planning involves engineers in the conversion process which, in turn, gives them a stake in the final outcome. The engineer's technical knowledge also allows planners to determine the capacity of a facility to convert to civilian production through a greater understanding of a facility's existing capabilities. The use of facility-level personnel is beneficial in two ways: (1) the engineers who are involved in the planning process become a knowledgeable technical work force when the facility is converted and (2) the cost of retraining and retooling is lowered because the conversion plan uses the skills and technical knowledge of existing personnel. Once a firm chooses a new product, engineers' efforts can be directed towards solving technical problems.

Regional Differences

Defense industries are concentrated geographically in the so-called gunbelt areas. These areas include especially California, Texas, and Massachusetts. Any level of conversion will affect these regions disproportionately. If defense cutbacks occur, many of the engineers in these areas will need to relocate to find employment. Relocation will compound the negative economic consequences for individual engineers if they have to sell their homes in a depressed market and pay expensive relocation costs. Older engineers require more time to find other employment, and they are likely to experience a heightened level of difficulty in adjusting to the conversion process.

Summary

The impact of economic conversion on engineers depends on the rate of economic conversion, the extent to which conversion is planned, and the extent to which it is decentralized. The concentration of defense engineers in particular specialties, firms, and geographical regions, suggests that, over the short term and particularly for some subgroups of engineers, loss of jobs and income is likely.

Over the long term and with good planning, the impact on engineers could be positive. If the economy de-emphasizes production of military hardware and diverts resources to producing large numbers of consumer products and building infrastructure, opportunities for engineers could be maintained and possibly expanded.

The Impact of Engineers on Conversion

A variety of common behavioral patterns, values, and cultural norms are held by engineers. Information about these characteristics provides some insight into the influence that engineers may have on economic conversion.

Political Orientation

Engineers tend to be members of conservative political organizations and to avoid political issues. Engineers like to be challenged by solvable, discrete problems. Consequently, the compromises associated with the solutions to social problems and the open-ended nature of those problems may be unappealing to engineers. Similarly, engineers, like many other professionals, have historically been uninterested in joining unions; this dislike of labor organizations may be a reflection of their conservative attitudes and of support for the status quo. Engineers who lack interest in social change will oppose economic conversion simply because of the change it represents.

Historically, company managers have chosen the problems on which engineers work. This has led most engineers to feel that they lack autonomy on the job. Managers' power can increase the value engineers place on conservative political orientations because work on high-technology projects that require security clearances is not likely to be available to those who are considered political risks. The structure of the workplace for engineers, like many other workers, creates work conditions in which they become passive and lack participation in policy making. If decisions about conversion are made solely by managers, engineers will remain passive recipients of social change. Conversion, however, offers a unique opportunity to alter the structure of the workplace by involving engineers in the planning and decision-making process; these changes toward greater participation may improve job satisfaction and productivity over the long term.

In general, engineers do not actively seek membership in political organizations that influence their value systems; however, many engineers are members of professional engineering societies. These societies have not attempted to influence engineers' values, but merely to reflect those values. Their role in economic conversion is likely to be primarily as a clearinghouse for information. On the other hand, conversion offers professional societies the opportunity to actively involve their members in debate and discussion about conversion, which can lead to the mobilization of the members and the emergence of leaders on this issue.

Values

Little research has been done on engineers' beliefs about justice or about their personal values, which together will influence their opinions about economic conversion. Both concepts deal with basic understandings about right and wrong. Justice is the application of what is viewed as right and wrong to the treatment of individuals in society, while values are personal definitions of right and wrong that guide individual behavior.

If engineers value conversion and believe it is just, they will be more likely to support the changes conversion will bring. Alternatively, if engineers do not value conversion and believe it treats them unfairly, they are unlikely to be supportive.

If engineers are to value the conversion process, they need to see a connection between the abstract goals of conversion, such as a stronger economy, and the effects of conversion on their work lives. This requires an articulation of the social alternatives or opportunities provided by conversion and the economic and social factors that argue for conversion, in language that engineers find understandable and appealing.

Economic conversion may require engineers to accept responsibilities such as discussions with end users, which they typically have not considered to be their own. Indeed, even some aspects of engineers' personalities, i.e., their need for order, desire for an unbroken path of career advancement, and their preference for conservative traditions, potentially conflict with the requirements of the conversion process.

A common belief among engineers is that absolute truths can be found if given enough time and effort for the search. This belief in absolutes often extends to their views about justice. Thus, engineers, who view justice as an absolute truth, find themselves in sharp contrast with the views of most social and behavioral scientists, who openly struggle with the implications of a variety of philosophical views of justice. These differences between perspectives can alienate engineers from economic conversion insofar as conversion carries relative rather than absolute value. Engineers who do not value the relativistic definitions of conversion may not be able to work with people who do; such a situation could impede the interdisciplinary approach needed for successful conversion planning.

Engineers have a deeply rooted optimism about the effects of technology, whether that technology is defense- or civilian-oriented. The focus of engineers is on the pursuit of technology, and many individuals attracted to engineering have placed lower value on social or political action than non-engineers. Indeed, many defense engineers believe that the products they design maintain world peace, and they believe that changing the existing system could potentially destabilize the world order.

Since engineers value technology's role in providing national defense, the changes caused by economic conversion may be seen as a threat to national security. This value conflict is a major hurdle for acceptance of the conversion process by engineers. If defense engineers are to value economic conversion, national defense must be redefined in terms of a strong national economy. If engineers do not value this broader definition of national security, their impact on economic conversion may not be positive.

Attitudes Towards Employment

Attitudes are more specific; their impact on motivation is more readily apparent than the impact of values. Some evidence now suggests that professionals in the U.S. are beginning to redefine the meaning of success. They are looking for meaningful work, in addition to well-paying work. Insofar as engineers keep their personal beliefs and feelings out of the workplace, they will probably be among the last professionals to participate in a redefinition of work. This reluctance to change will inhibit engineers' support for the conversion process.

Resistance to change in the current system is apparent in the majority of defense engineers whose opinions indicate that they would be proud to work on weaponry (especially if it is non-nuclear). However, as greater numbers of women and minorities enter the profession, engineers' attitudes towards defense work could change. These groups usually place less value on the development of defense-related products.

Summary

The value systems and attitudes common to most engineers indicate strong support for the existing employment system. Economic conversion represents a radical change in the engineering workplace. The negative consequences of economic conversion on defense engineers and the need for a significant shift in values for successful conversion lead to the conclusion that engineers and engineering societies should not be expected to be dynamic advocates for economic conversion or to provide leadership in planning economic conversion. Historically, engineers have not responded to the political implications of their work. Neither dynamic advocacy in favor of conversion nor open rebellion against economic conversion is likely. Engineers can be expected to tackle whatever problems society, or their companies, pose without making value judgments about those problems.

Factors that Could Increase the Participation of Engineers

The factors that may encourage engineers to participate actively in the conversion process are, in part, a consequence of these impacts. These factors, however, are poorly understood. Several were suggested at the workshop. Most importantly, it is clear that economic conversion will be more successful and less disruptive if engineers are involved.

Understanding the Issues

Engineers are pragmatic and can be moved to action if they believe that the political and economic environment of the nation, and of the world at large, is changing. The availability of information about the economic, social, political, and moral issues involved in the conversion process, in addition to the more obvious technical issues, could raise the level of meaningful discussion among engineers. Meaningful discussion, in turn, might increase participation while minimizing negative consequences and facilitating a smoother conversion process. Access to accurate and up-to-date information is essential if engineers are to participate in economic conversion. Professional societies could play a very important role in developing and disseminating such information.

Personal Resources

The time and effort needed to learn new engineering skills and to change work routines at the end of a career will be difficult for many older defense engineers. If managers decide that retraining older engineers is not cost-effective, conversion will be particularly difficult for these engineers. This management decision is likely for two reasons. First, companies may be unwilling to pay large salaries that reflect years of engineering experience. Second, managers will prefer younger engineers with longer work lives if they believe that older engineers only have a limited time to contribute to the company. As a result of formidable obstacles to re-employment, many older engineers will see early retirement and/or nonengineering work as their only alternatives.

Access to Continuing Education

Continuing education can be an effective approach to preparing engineers for a changed job market. Continuing education can provide the retraining that defense engineers need for employment in the civilian workplace; it can also enhance interest in the technical problems associated with civilian engineering jobs. Providing defense engineers with new skills ensures that society does not lose the valuable technical skills defense engineers can bring to civilian jobs. To transfer current engineers to alternative employment areas such as hazardous waste cleanup, landfill design, medical technology, and engineering software technology will require retraining to develop new skills.

The Extent of Planning

Where planned conversion occurs, personal resources may not be as critical to participation because less time and money are needed to find new jobs. Planning would involve engineers in the identification of the civilian products that defense plants can produce, as well as the development of a detailed plan for converting to consumer production, retraining employees, and determining market feasibility. An example of this planned process was the conversion of machinery in the aerospace industry to build mass transit systems.

Successful planning should include implementation of national policies that will bridge the cultural gap between military and civilian engineering and help make the transfer from the military sector easier for engineers. The national laboratories are already attempting to formulate guidelines and communication links between corporate needs and lab capacities. The successful conversion of the labs is important because it could provide engineers with examples of successful transfer from military to civilian goals. The successful transfer of technology from the labs to the private sector can convince defense engineers of the feasibility of economic conversion.

Successful planning should also match the skill of engineers in the defense sector to the opportunities available in the civilian sector. Such "matches" must provide opportunities to design and manufacture products that provide engineers with interesting and challenging problems.

Changing Nature of the Workplace

New management techniques and a philosophy of worker involvement are changing the nature of the workplace. These changes range from total quality management (TQM) and quality circles (QC) to workplace democracy and employee stock ownership plans (ESOP). Although motivated in part by desires to increase worker productivity and international competitiveness, these changes in the workplace offer opportunities for involving engineers and technical workers in economic conversion. Greater autonomy and decision making in the workplace will be a source of job satisfaction that will make jobs in converted plants and industries more desirable than defense work that does not seek to actively involve workers in management.

Industrial Policy

The federal government can contribute to an overall conversion policy by designing an industrial policy that targets the goals of conversion. In general, industrial policy is considered a set of actions taken by government aimed at stimulating industrial development and, in the modern context, competitiveness in the international marketplace. Implementing an industrial policy is accomplished by providing companies with adequate capital, educated work forces, R&D support, and a variety of international trade incentives and barriers aimed at assuring access for a country's products to the market and in some instances limiting competing nation's products in its domestic market.

Because the cost of developing technology is prohibitive and the federal government has limited resources, the most important element of industrial policy is the selection of industries for government support. In the case of economic conversion, industrial policy supporting conversion would select future consumer-oriented technologies and commit resources to transferring the technologies, skills, and work force from defense industries to these areas.

International Competition

Manufacturing competition from Europe and the Pacific rim could provide a major impetus for engineering participation in economic conversion. Foreign competition affects engineers in two ways: it redirects the goals of U.S. industry toward the development of internationally competitive consumer products and it redirects the focus of engineers from "defeating the communists" to "beating" international companies that are trying to invade U.S. markets. Defense engineers are likely to respond as positively to the new international challenges as they did to the old challenges, although this scenario is mitigated by the growing number of multinational corporations.

Summary

Engineers' participation in the economic conversion process will depend on their knowledge about the technical dimensions of conversion and about the job market, the flexibility of their engineering skills in transferring from the defense to the civilian sector, opportunities for continuing education, the level of government and corporate planning that accompanies the conversion process, and engineers' personal resources for relocation and retraining purposes. Women, minorities, and the new generation of engineering students will be increasingly interested in nondefense work. The contributions of these groups offer a stimulus to the participation of engineers in the conversion process.

12
Public Policy Directions

Martha W. Gilliland, Patricia L. MacCorquodale, and
Jeffrey P. Kash

A full public policy analysis was beyond the scope of this project. Clearly, however, the multidisciplinary perspectives offered in this book on the interface between economic conversion and engineers identify conversion as a complex public policy issue. In order to design a comprehensive public policy that targets the needs of engineers and the goals of economic conversion, perspectives discussed in this volume need to be addressed. This chapter only begins the discussion by identifying some public policy directions. These directions are not specific enough to be termed policy options and they have not been analyzed for their costs and benefits.

A core question is one of how to empower engineers and engineering societies to participate in, support, and provide leadership for the nation's economic conversion. Thus, the first section of this chapter summarizes components of empowerment based on the information about engineers provided by the multidisciplinary perspectives. The second section identifies activities of key institutions that could facilitate that empowerment and encourage participation of engineers. The final section presents components of a research agenda on conversion.

Providing the Personal Connection: Empowerment

Engineers will become involved in and actively support economic conversion to the extent that they feel personally connected to the process and empowered to participate in "making it happen." Empowerment depends on a sense of interpersonal efficacy and social responsibility; it requires a sense of autonomy and trust in one's own abilities. For engineers it requires involvement in the issues, opportunities to be creative, technical challenges, options, respect, and economic reward.

Consequently, public policy aimed at empowering engineers will have to provide at least four outcomes.

Involvement in the Debate

Creating a personal connection requires participation in the debate about conversion on the job, in the community, and at the state and national levels. For example, to the extent engineers speak to the public about conversion, both the public and the engineers increase their participation in the debate. Within industry, seeking the involvement of engineers in management decisions about new directions for their company may also foster greater participation. The more involved engineers become in the conversion debate, the greater are the opportunities for successful conversion.

Opportunities to Apply Skills and Creativity

The opportunity for self-expression through applying skills is also a key element of empowerment. Applying skills in new directions individually and as part of teams generates creativity. Engineers have technical skills that have been directed at defense-related technologies. Their support for conversion requires that they be provided with specific technical challenges in the civilian sector for such priorities as infrastructure improvement, environmental protection, economic competitiveness in manufacturing, improvements in the standard of living, and so on.

An Appeal to Values

Engineers' codes of ethics offer an important arena for connecting to values. For example, most of the codes begin with the statement: "engineers must be dedicated to the protection of the public health, safety, and welfare and recognize that their work has a direct and vital impact on the quality of life for all people." One code states: "engineers consider the consequences of their work and societal issues pertinent to it and seek to extend public understanding of those relationships."[1]

Respect and Economic Rewards

Engineers, like other workers, find job satisfaction from the material rewards of their employment. Many engineers see defense work as attractive because they believe it is better paying, more prestigious, and more interesting than civilian work. Providing job satisfaction in the commercial sector requires attention to comparable salaries and prestige.

Key Institutions

Federal Government

The federal government is the most important institution for promoting conversion policy. The federal government influences conversion policy in four ways: (1) it controls vast financial resources, which can be used to support conversion policy; (2) it can disseminate information supporting conversion policy to all levels of society through government agencies; (3) it can pass legislation that provides incentives for businesses and individuals to support conversion policy; and (4) it controls defense expenditures and can help initiate conversion by diverting defense funds. Five types of federal government activities in support of conversion are identified.

Plan the Conversion

This study has documented the need for conversion planning; with planning, possible negative impacts will be less severe and of shorter duration. Reductions in military spending represent reductions in federal spending. Consequently, it is the federal government, through Congress and the Executive Branch, that must provide leadership and resources for such planning. As noted in Chapter One, numerous bills that define the focus of the shift and provide resources for planning and retraining have been introduced in Congress. None has become law. Some consensus on the direction of the shift (e.g., infrastructure, manufacturing, environment) is required, and that specificity must come from the federal government.

Consensus will be difficult; the existing military-industrial complex has a natural resistance to conversion simply because of the redistribution of government resources conversion represents. Consensus within Congress alone will be equally difficult. For any major policy area, there is a coalition of members of Congress, bureaucrats in the executive branch, and special interest groups who support that policy. Powerful members of Congress are on committees that control defense spending. Interest groups provide political support to members of Congress in exchange for favorable policies. Such groups also provide political support to bureaucracies by promoting the continued authorization of programs that benefit them. In the parlance of the political scientists, this is known as the iron triangle. Interest groups support members of Congress who, in turn, authorize desired programs. A bureaucracy implements the programs for the benefit of the interest group, which, in turn, supports the bureaucracy before Congress.

As applied to defense policy, members of Congress who sit on committees overseeing the Pentagon, such as the Armed Services Committees and the Appropriations Committees, are typically from states that are economically heavily dependent upon defense spending. Defense contractors provide extensive support to such members in exchange for beneficial programs, e.g., the authorization of weapons programs. Members of the Pentagon, who are charged with overseeing weapons development, direct that such programs be built or based in the home states or districts of the critical members of Congress. All three actors in the defense iron triangle have incentives to develop, build, and deploy weapons systems. No one actor faces incentives to cut defense spending. Many members of Congress will aggressively attempt to protect the defense programs in their states. Thus, substantial military spending cuts that would act as a stimulus for conversion will be difficult to bring about. Consensus on the direction of the shift will be even more difficult. In short, planning is needed but it requires consensus about the extent of cuts and the focus of the shift, both difficult tasks.

Refocus the Priorities of the National Laboratories

The multiprogram national laboratories that are funded through the U.S. Department of Energy conduct research on technologies with applications in defense. This includes the design, development, and testing of weapons systems. These multiprogram laboratories include Argonne, Brookhaven, Lawrence Berkeley, Oak Ridge, Pacific Northwest, Idaho Engineering, Lawrence Livermore, Los Alamos, and Sandia national laboratories. Total budgets for all of these laboratories have been in the range of $5 billion to $10 billion per year.

The laboratories are immersed in the defense engineering culture; if they can shift to research and development of products for the civilian sector they can model the elements of such a shift for the private sector. More specifically, much of the engineering work in the labs is very "high tech" and almost entirely performance-driven. Shifting this engineering culture to consider cost, consumers, and the environmental compatibility of technology is a major challenge. The labs, in fact, may be pivotal in demonstrating how defense engineers can become civilian engineers. If the labs can be managed to bring about the shift, in so doing they may provide new technology and a model for defense companies in the private sector. Success may, in fact, determine whether the labs continue to exist in the face of declines in the overall defense budget.

The shift from defense-related research to research on technologies with commercial potential is a fundamental change. In addition to a change in the engineering culture, it requires strong linkages between the labs and private companies, linkages directed at providing jobs for engineers and at developing new and improved technologies for solving problems in the public domain (e.g., infrastructure and environmental protection). To date, promotion of the capabilities of the national labs by groups, such as the Interagency Committee for Federal Laboratory Technology Transfer, have resulted in an increase in cooperative research and development agreements between the federal labs and the private sector. Between 1987 and 1990, such agreements increased by ten times.

Even more encouraging, in July 1992, the Senate passed the Department of Energy Laboratory Partnership Act (S. 2566). The bill directs the secretary of the Department of Energy to ensure that the labs enter partnerships with private companies and educational institutions to develop technology in critical areas such as energy efficiency, energy supply, high-performance computing, environmental protection, advanced manufacturing, advanced materials, and transportation. The bill requires that mechanisms for evaluation of the partnerships be developed and that incentives for industry and educational institutions to participate in the partnerships be formulated.

Establish Joint Ventures

Joint ventures involving federal laboratories, universities, and private companies and directed at developing technology for solving problems in the public domain can be powerful. R&D programs designed to address critical social problems while providing new consumer products at the same time are possible. One example is the integration of pollution-prevention technologies and advanced computer-aided manufacturing systems.

Provide Direct Support for Retraining

In addition to the indirect support for retraining that is associated with re-focusing the national labs and establishing joint ventures, direct federal support for retraining engineers is also needed. Fellowships for engineers to update their skills at universities are one approach. Funding in support of retraining programs developed by the professional societies is another.

Provide Incentives to Defense Companies

The federal government can offer incentives, such as investment tax credits, to defense companies for the implementation of innovative programs that focus on the development of technologies to solve societal problems. Obviously, such incentives raise the issues of government interference in the marketplace. However, options in this category ought to be explored. The vulnerability of small companies, with limited financial resources, to business cycles could be reduced by encouraging cooperation and risk sharing in R&D with other companies and with government.

Engineering Societies

The professional engineering societies are in an excellent position to provide leadership for their members. Most engineers are members of one or more professional engineering societies (see Appendix B for a list of engineering societies); engineers generally read their society's publications; and they are generally active in local chapter meetings. Substantial numbers also attend their society's national meetings. Moreover, the societies have established communication networks; they have administrative support; and they have established legitimacy from the perspective of the public and political leaders.

The societies can:

•involve the membership in the conversion debate and in discussions about its directions and impacts through articles in the journals, presentations at local chapter meetings, and forums at national meetings.

•access, summarize, and distribute information about conversion from Congressional staff and the national organizations that are now devoted to conversion.

•design and implement retraining efforts.

•lobby for Congressional bills on conversion that provide assistance to engineers.

•involve the membership in discussions of the Engineering Codes of Ethics as they may apply to conversion.

•identify specific technical challenges associated with solving problems in the civilian sector such as those inherent in environmental protection globally, nationally, and locally.

•serve as a clearinghouse for identifying and communicating retraining opportunities.

State and Local Government

State and local governments can mitigate the impacts of declines in defense spending within their borders by working in partnership with defense companies on conversion of specific factories and military bases within their boundaries. They can also apply for federal monies and distribute these funds where they are needed within the state. This has been the focus of efforts at the state and local levels in Texas, Arizona, and California.

Industry

Industry itself probably plays the most important role in educating working engineers. However, in order to retrain successfully and remain competitive, individual corporations must first identify the civilian products on which they are going to focus. One defense company with an aggressive approach to conversion toward commercial production is Hughes Aircraft. Hughes defines the differences in market development between defense and commercial markets as the need to: delineate requirements for products internally (unlike defense production where military specifications define products), be first to enter commercial markets, design systems that produce payoffs for customers, understand commercial markets and customers, and select markets in which Hughes has unique sustainable competitive advantages (e.g., markets where Hughes owns patents in important technologies).

The technologies that Hughes is developing for commercial use are based on defense technologies. Examples follow.

Optical Character Recognition Systems

This technology is based on artificial nerve networks and "fuzzy" logic developed in work on smart missile systems. The technology recognizes patterns such as zip codes or writing on documents. Hughes is now developing a system using optical character recognition under a contract with the U.S. Postal Service.

Smart Highways

This highway monitoring system uses cameras and computers to monitor the types of vehicles that travel on roads and highways; it utilizes video image processing. The information from video cameras is fed to a system that notifies a human controller if a traffic situation needs attention. The system is unique because it relies on a limited number of operators; the system itself can distinguish between routine traffic and traffic problems.

Smart Cars

In conjunction with General Motors, "heads-up displays" (HUD) for cars are under development. The system is based on technology developed for fighter aircraft. In aircraft, HUD allows the pilot to monitor the plane's functions without looking down at the controls; an image of the control panel is displayed in the plane's canopy, so that the pilot can continually scan for enemy aircraft. In automobiles, HUD technology is used in cars for similar purposes; drivers can assess their speed without taking their eyes off the road. The technology has been successfully applied in two models of GM cars currently in production.

Electric Cars

Hughes also developed an inverter box that converts direct current to alternating current. This is used in GM's electric car program to run the electric motors for prototype electric vehicles.

Automated Manufacturing Systems

Three products that will help increase efficiency in future production lines have been developed. First, the open-architecture controller allows manufacturers to change ongoing processes on the factory floor quickly. Second, a real-time hardware-software integration system allows manufacturing software to adjust hardware problems to changing conditions on the manufacturing floor. Third, defects and product quality can be assessed using a noise-monitoring system. This system listens for excess noise in products at the end of the manufacturing process. The technology is based on digital signal processing techniques learned from defense work on sonar systems.

Satellite Network Systems

Individual satellite networks that connect organizations with affiliates anywhere in the world are now being marketed. These VSAT networks allow companies to communicate information about inventory, credit applications, and financial transactions among members. Hughes has sold these systems to Circuit City, GM, Wal-Mart, Holiday Inn, and the People's Republic of China.

Liquid Crystal Light Valve Projectors

Now in their final stages of development, the systems amplify white light to produce high-definition images that may be used in business presentations or in theaters.

Both the conversion process at the national level and the industries themselves at the corporate level would benefit from involving the working engineers in planning the conversion at the corporate level. Involvement empowers and yields high productivity and creativity. Total quality management strategies as they are being adapted by industry offer one mechanism for involving engineers in a team approach. TQM asks: (1) who is the customer, (2) what does the customer want, and (3) how do we build reliable products for that customer? Teams involve engineers, salespersons, assembly workers, and management.

Universities

Education is a key element in successful economic conversion. Over the short term, educational efforts must target unemployed defense engineers for retraining. Over the long term, efforts must target future engineers who will require new skills to obtain jobs in a different kind of economy. For both types of efforts, cooperation among government, industry, and universities is needed; retraining has been identified as one of the activities in which both government and industry must become involved. Universities also have a key role.

Evaluation of the current approach to undergraduate engineering has been underway for several years by educators and the engineering accrediting organization. Issues associated with economic conversion simply provide additional impetus to this ongoing effort. Traditional disciplinary boundaries within engineering are of questionable utility, since technology development now draws on many of the subspecialties simultaneously. Similarly, the world of technology development in the private sector is moving toward a team approach, as total quality management strategies are adopted. Companies want to hire engineers that have learned and experienced TQM in their undergraduate curriculum. The companies that hire engineers have long argued that engineers need a broader education, one that allows them to develop interpersonal skills to be effective in group work and one that forces them to obtain good, if not excellent, oral and written communication skills. The current technical focus of engineering education reinforces work at the individual level and problem solving, at the expense of team work and communication abilities.

Supporters of change in the current system believe student engineers should be exposed to a wide variety of educational disciplines, with less emphasis on the technical. M.L.A. Mac Vicar suggests "a true educational partnership among the technical, arts, social and humanistic disciplines so that on some level students see the interrelationships between science and technology on the one hand, and societal, political, and ethical forces on the other."[2] This approach would increase engineers' awareness of the rest of the world and encourage them to see the social impacts of their work. This new awareness would enhance newly trained engineers' ability to see beyond the technical nature of their work.

The expansion of engineering education to include a multidisciplinary education would change the focus of engineers. E. Wenk believes that in future engineering "the focus must shift from preparation for the design of elements or systems of hardware, to understanding and design of complex technological delivery systems that also incorporate public and private institutions and communication networks with social processes and the cultural preference to produce goods and services for a world community in the 21st century."[3]

A Research Agenda

Better information about economic conversion—its impacts and how to mitigate them—would provide for more effective public policy. For example, conversion policy will be better if the following questions are answered: how many engineers are employed in the defense sector, what are the projected changes in the engineering work force (e.g., the entrance of women and minorities), and what type of engineering skills do defense engineers have. The answers to these question help determine the size of training programs, the amount of funding that is needed, and the amount of involvement needed from industry and government. An evaluation of case histories on conversion, such as the conversion from military to civilian production in Europe after World War II, could suggest better policy for the 1990s. In addition, further surveys about engineers' attitudes and values would provide insight into how to design a conversion policy that motivates engineers.

The need to address three related questions drove the development of this research agenda.

(1) What do we need to know to plan for a smooth economic conversion?
(2) What do we need to know to evaluate the impact of various economic conversion scenarios?
(3) What do we need to know to be able to project patterns of displacement of defense workers as defense expenditures decline?

These questions and the varied research approaches that they suggest generate research agendas with varying time frames. Specific research issues that will help answer these questions under three time frames are discussed: (1) those that can be addressed relatively easily in the short term, (2) those that require a multiyear effort, and (3) those that focus on more fundamental restructuring and would benefit from a long-term approach.

An Applied Pragmatic Research Agenda for the Short Term

There is an immediate need for better data on two dimensions of economic conversion: (1) defense spending and (2) engineers. These data need to be gathered systematically using agreed-upon definitions to establish baseline information for use in making projections about the magnitude and effects of economic conversion. Clearly, understanding the extent to which the nation's pool of technological talent is engaged in military-related activity is important to understanding both the economic impacts of this particular use of resources and the nature and extent of the technologist conversion problem.

Unfortunately, these data are not readily available. In the past, data on engineering and scientific employment by area of critical national interest (from the NSF, for example) have combined data categories, making it difficult to segregate military use from other employment. In addition, the data have suffered from unduly narrow definitions of what constitutes military-related activity or used inappropriate methods of estimation. For example, in 1981, the Department of Defense estimated that only 12% of the nation's engineers and scientists were engaged in "defense-induced" employment. Later in the 1980s, the NSF released an apparently corroborating estimate based on a sample survey of engineers and scientists. Yet the method used to develop the DoD estimate explicitly excluded "defense-induced" employment related to arms exports, along with the military-related part of the space program and the entire program of research, design, testing, and production of nuclear weapons. The method also assumed that the percentage of engineers and scientists in any given industry is the same in the military-serving part of the industry as in the civilian-oriented part, an unrealistic assumption for whole categories of critical military industry (such as aerospace). This approach guaranteed that DoD estimates would be far too small, casting doubt on the accuracy of the NSF survey estimate as well.

The central point is that there has been too much "playing around" with what has become an increasingly critical set of data, perhaps because it has been a very "political number" due to the inordinately large amount of the federal budget spent on defense relative to other government spending programs. Pragmatism now requires that all considerations of "making the number come out right,"whether on the part of supporters or critics of the military sector should be put aside. We must understand where we are before we can effectively determine how we will get where we want to go.

More than in most areas of human endeavor, the rate of progress of technology is very strongly dependent on the activities of the "best and the brightest." Whatever the real fraction of engineers and scientists employed in military-related work, the impacts of that use will be radically different if the better salaries and facilities available in the military sector have taken a disproportionate number of the most productive technologists. To the extent that it is possible to gather data on quality as well, that path would certainly be worth pursuing.

A Technology Assessment of Economic Conversion for the Midterm

The approach of technology assessment can be applied to the issue of economic conversion. The basic goal of technology assessment in this case would be to project patterns of displacement of scientists and engineers in specific fields as military expenditures decrease. From these projections policy makers could develop appropriate public policies that could respond to these patterns.

A technology assessment requires that scenarios of economic conversion vary along at least two dimensions: (1) the rate and level of conversion and (2) the priority outcomes from conversion. The goals of the assessment would be to specify boundary conditions for these dimensions and then to build a comprehensive model to generate three to five alternative scenarios. Through these scenarios, the rate and priority outcomes of economic conversion can be varied.

With respect to *rate*, the boundary conditions for speed need to be specified. A rapid conversion scenario, for example, would be a situation where 90% of the military budget is converted over 9 years (at the rate of 10% per year). The slow-paced conversion, in contrast, would include no major cuts in the military budget (e.g., 1%-2% reduction per year over 9 years). The boundary conditions would then specify, given the figures on the current situation, the dollar amounts under the fast and slow case scenarios and the number of military personnel and engineers who would be affected.

For the alternative scenarios, rate of conversion, for example, could be set at 25%, 50%, and 75% expenditure (real dollar) reduction in 10 years. In order to analyze the impacts of economic conversion on engineers, the model should consider the structure of the conversion within the defense sector. For example, characteristics of engineers (their training and specialization, numbers, age structure), and distribution of cuts between labor (armed forces) vis-a-vis weapons would be included. The scenario can also be examined for its impact at the regional, national, or industry level.

Finally, the models will assess the *priority outcomes* of conversion, i.e., how will we spend the money saved through defense cuts? Not only could variation in the degree of planning for conversion be assessed, but assessments could be made of the effects of priority outcomes. Five areas for additional spending that could benefit from conversion were identified in this book. Some participants believed that manufacturing competitiveness was the highest priority (see Kent Hughes on technical priorities and Dumas on economically contributive goods and services). Others called for investment in infrastructure maintenance and rehabilitation (Vlachos), environmental challenges (Vlachos, Yudken and Markusen) energy, and social programs (Yudken and Markusen).

In a technology assessment, each of these scenarios can be analyzed in order to determine: (1) specific employment impacts for engineers by specialty and (2) the general implications for the economy (GNP, debt, trade balance). The results from modeling are artificial constructs that may not reflect the complexity of actual economic conversion; however the different scenarios provide a starting point for policy makers. The information provided by the models can be used as a baseline for defining the effects of conversion on society. By identifying some of the results of conversion, models allow public policy to mitigate some of the possible negative consequences.

The Long-Term Approach

The ability to understand the impacts of and to plan optimally and systematically for economic conversion would be greatly enhanced by research on at least ten issues. These are listed here along with an indication of the utility of the information provided. Research on all of these items would require substantial time and resources.

Implications of Planned Versus Unplanned Conversion

This issue emerged as one of the central themes of the workshop. Some participants believed strongly in the competitiveness of the "free-market" system, in which conversion involved only a decision on the size and rate of cuts in military spending. Market forces would then take care of allocating these resources into specific sectors or products and no extensive planning would be needed. Yudken and Markusen argued that "ordinary market forces will not be sufficient to help scientists and engineers make the transition from military to civilian-oriented jobs." Jameton and Vlachos argue planned conversion be framed in a global perspective directed at understanding the global impact of reduced arms expenditures on national and international economies and justice.

Historical Reconstructions of Economic Conversion in the United States, Great Britain, and France

Economic conversion is not a new phenomenon. The experiences of the United States, Europe after World War II, and current efforts in Eastern Europe and the former Soviet Union may be particularly valuable in identifying factors that ease the conversion transition, improve the economy, and contribute to other social goals.

Case Studies of Communities Affected by Base Closings and of Firms that Have Converted

The Office of Economic Adjustment has been working with communities affected by base closures since 1961. Case studies of the communities with which that office has worked would be useful in identifying the regional and community parameters associated with conversion. Many firms have experienced varying levels of conversion. For example, Kaman Corporation reoriented from helicopters to guitars; and Hughes Aircraft currently is undergoing a major conversion effort. Case studies of firms would be particularly interesting insofar as they reveal the involvement of engineers in the conversion process.

Surveys and Longitudinal Studies of the Values, Attitudes and Career Paths
of Engineers

Although there is widespread agreement that engineers can play a valuable
role in economic conversion planning, little is known about engineers'
values and attitudes (Everett and Gagné). Issues to be addressed include:

What is the impact of work experience, industry, life-cycle position,
and political climate on engineers' values and attitudes?

If engineers are fascinated with technology, why are they less
satisfied with their jobs over time and is this fascination related to
understanding the human impact of technology?

How does the mobility of young engineers occur when there is little
shifting between defense and civilian sectors?

A Comparative Analysis of the Skills and Experiences of Engineers in the
Defense Sector to Those of Engineers in the Civilian Sector

This research would focus on the issue of transferability of skills across
sectors (Dumas, Yudken, and Markusen). Of interest is whether defense
engineers are more specialized, higher paid, and use systems analysis and
design more extensively than their civilian counterparts. If the engineering
in the defense sector is oriented toward maximum performance rather than
cost, assessments are needed as to the difficulty of changing this orientation
and what retraining for engineers it implies. Finally, an inventory of skills of
defense engineers is needed to assess civilian applications of these skills.
This inventory may be especially important for engineers who work in
sectors that are heavily defense-dependent, e.g., aeronautical engineers.

Case Studies of Engineers who have Moved from the Military to the
Civilian Sector

Because engineers can play a leadership role in economic conversion, case
studies of engineers who have moved from the military to the civilian sector
are needed. Identification of the values, attitudes, and life-cycle position of
these engineers would be useful in locating engineers who can play a
leadership role in their firms and engineering societies. An assessment of
the coping strategies used by engineers and difficulties encountered during
job transitions would provide essential information for designing assistance,
retraining, and education programs to aid engineers affected by conversion.

Case Studies of National Laboratories

National laboratories have recently begun technology transfer programs.
Information on why these programs are or are not working would be helpful.
For example, numerous labs are applying their expertise to hazardous waste
disposal and reclamation of natural resources from these wastes.

Surveys of the Activities of the Engineering Professional Societies Related to the Issue of Economic Conversion

Included should be activities ongoing now and during times of major past conversions (e.g., post-World War II).

Philosophical Issues

Little is known about the values of engineers, if and how those values are related to their work, and how those values will be affected by shifting personnel from military to civilian projects.

Sociological Issues

Conversion offers a unique opportunity to study workplace democracy and the effects of greater participation by workers in decision making when engineers are involved in the conversion planning and implementation processes. Economic conversion suggests significant changes in the workplace. For example, what will be the effects of less secrecy, less specialization, and greater participation in decision making upon worker satisfaction and productivity.

An Economic Conversion Institute

To conduct the research on our agenda, we propose the formation of an Economic Conversion Institute at a major university to conduct the technology assessment of economic conversion, initiate long-term research, and act as a clearinghouse for information and data concerning economic conversion. This institute would provide an interdisciplinary forum where scholars and practitioners from a variety of disciplines focus on the long- and short-term implications of economic conversion. Because input from the engineering sector is essential to the success of such an institute, the institute should be established as a partnership between academia, industry, and engineering societies and should actively facilitate communication and collaboration between engineers, scholars, planners, educators, and policy makers from all sectors. An institute specifically charged with economic conversion can cooperate with government agencies, such as the Office of Technology Assessment, to insure that their research efforts include economic conversion and to disseminate and synthesize research results from a variety of sources.

Notes

[1]*Model Guide for Professional Conduct*, American Association of Engineering Societies.

[2]M.L.A. Mac Vicar, 1987, "General Education for Scientists and Engineers: Current Issues and Challenges," *Bulletin Science Technology and Society* 7, No.5 (1987): 592-597, as quoted in Joseph R. Herkert, "Science Technology and Society Education for Engineers," *IEEE Technology and Society Magazine*, 1990: 22-26.

[3]Edward Wenk, "Social, Economic, and Political Change: Portents for Curricular Reform in Engineering Education," Session 3264 in Lawrence P. Grayson and Joseph M. Biedenbach (eds.), *1988 American Society for Engineering Education Annual Conference Proceedings* 4&5 (1988): 1724.

Appendix A
The Participatory Research Process

The goal of the workshop was to access the collective wisdom of the group by creating a highly participatory, nonjudgmental, collaborative atmosphere. The synergistic effects of such a group process produce a high level of analysis. The workshop design was guided by the belief that a participative format was essential to (1) expression of a full range of beliefs and (2) development of a plan of action for further research, policy, and professional organizations.

Specific guidelines and ground rules were enunciated to all participants. The workshop was designed to separate idea generation from idea evaluation, to encourage dialogue, and to make assumptions clear. The orientation was problem-solving with a specific goal to synthesize knowledge and struggle to apply it to economic conversion. Participants were asked to focus on problems to be solved and to assume that some level of agreement existed and could be strengthened as consensus emerged. Consensus was built by identifying likely and important high-priority outcomes. Agreement on some of the goals of society allowed us to work toward a common end in prioritizing the ideas that had been generated.

The highly participatory process required flexibility on the part of the organizers to respond to unanticipated directions in discussions. A refocus on nascent topics that looked worthwhile occurred throughout.

Diversity of Participants

Diversity of perspectives was assured by the selection of those invited to attend the workshop. Participants represented (1) scholars from a variety of disciplines, (2) various types of engineers, (3) employment sectors ranging from the defense industry to the university, and (4) several professional organizations, including the American Engineers for Social Responsibility.

Questionnaires were sent in advance to identify participants' orientations toward economic conversion and areas of agreement and disagreement. The survey included five questions.

(1) What is your view of the current status of economic conversion given current geopolitical events such as the Gulf War and the end of the Cold War?
(2) What would you like to see as outcomes of this workshop in order for you to feel that it had been successful and worth your time?

(3) Summarize two or three key issues that you feel are critical for the group to address.
(4) Do you have any thoughts on specific ways we can create a collaborative problem-solving atmosphere and structure?
(5) Ultimately, we would like to see specific results from our workshop, including the distribution of this new knowledge to policy makers, engineering educators, and professional organizations. At this point, do you have any ideas on how we can move ahead toward successful dissemination of results?

Responses were obtained from all participants either in writing or by phone. The results were used by the facilitators in assessing the degree of fit between the workshop goals and the design and orientation of the participants (questions 2, 4, and 5). Participants shared many of the goals of the workshop in terms of knowledge, the development of a broad, interdisciplinary perspective, a participative process, and the desire for networking. Respondents varied in terms of the action they foresaw as resulting from the workshop, reflective of the diverse sectors they represented. Nearly everyone agreed on disseminating the results of the workshop, identifying areas for further research, and recommending changes in engineering education. Smaller groups wanted to apply knowledge from the workshop to the implementation of economic conversion or to affect public policy. Others wanted to mobilize engineers or the public to encourage economic conversion; others were looking for direction with respect to professional associations, especially those from the American Engineers for Social Responsibility.

With respect to the substantive questions (1 and 3), opinions varied greatly. Some of the questions and issues raised would be answered in the papers and discussion at the workshop; others required additional research and, in fact, initiated the discussion of research agenda; and others extended beyond the scope of this project. Respondents were divided over the effects of current geopolitical events on economic conversion. The smallest group (26%) saw economic conversion continuing or accelerating. Their opinions were that "conversion will happen," " economic conversion is progressing at the expected rate of 2% per year," or "military downsizing is already under way." Others saw larger economic forces independent of current geopolitical events that were driving economic conversion. For some, these forces were market forces: "economic conversion will be a business decision to sustain profitability and survival." Others saw economic conversion as linked to the necessity of using technical resources for other pressing problems such as energy alternatives, resolving environmental issues and addressing medical problems.

A larger group saw recent geopolitical events as stagnating or reversing the conversion process. For some, the Gulf War and its demonstration of the high-technology military would "slow the pace of military cuts" or "encourage U.S. dependency on military sales." Others believed that the Gulf War "distracted attention from domestic needs" and represented increasing militarization of international politics. Others pointed to the underlying values evidenced by the Gulf War, values that are counter to those they believed necessary for economic conversion: "The Gulf War showed decreased respect by Americans for human life;" "the U.S. has decided to achieve its aim by force rather than negotiation;" "peace does not occur because war ends." Others believed that "people will avoid economic conversion until public and business adopt different value systems."

The remaining respondents (39%) believed that the effects of recent events were mixed. On the one hand, they agreed with those who foresaw stagnation in seeing the Gulf War as an arena that highlighted new weapons and technology. But they believed this weapons display might slow down military spending; for example, they saw the war as indicating that "the U.S. is over-equipped," there will be a shift from strategic weapons and large forces to tactical weapons and special forces, or the "visibility of arms merchants might create momentum for limits on arms exports." Others believed that the war and current events made economic conversion "now all the more important," but they lamented that "neither the public or private sector has a strategy for reducing the pain of economic conversion or capitalizing on its opportunities." Others believed that neither Congress, the Administration, the Department of Defense, nor the defense industry "had thought through the impact of post-war needs on the industrial base."

Respondents' identification of key issues revealed even more clearly the diversity of positions with respect to economic conversion. The responses were categorized in four general areas: the economic conversion process, forces behind conversion, outcomes of conversion, and aids to the process of conversion. With respect to the process, some respondents raised basic questions, such as what is economic conversion or why economic conversion, although these issues had been addressed in the project proposal and papers. With respect to the forces behind conversion, some wondered about the "level the U.S. should maintain to react successfully to future aggression," while others asked "why/how was the peace dividend subverted?" Some saw conversion as tied to the "promotion of global wealth and enhanced U.S. competitiveness" while others believed it was part of "the coming necessity to plan all economic activity within the framework of a sustainable society/environment."

As in other matters of beliefs, many participants' views were based on taken-for-granted assumptions and implicit values. Responses indicated conflicting and strongly held orientations. Because the workshop time constraints did not permit an exercise in value clarification and consensus building with respect to basic assumptions, the facilitators identified a list of assumptions that were embedded in questionnaire responses. This served to sensitize participants to the idea that their values and beliefs might not be shared by others. The following areas of disagreement were identified:

Economic Conversion

Economic conversion is of extreme importance to the engineering profession.

Economic conversion will happen.

Economic conversion creates regional economic challenges.

Engineers will be dislocated by economic conversion.

Economic conversion requires planning.

Engineers' Role

Dislocated engineers will need help in finding new jobs.

Engineers are conservative and will not get involved in economic conversion planning or politics.

Engineers should take a position on economic conversion.

Engineers should feel responsible for their role in economic conversion.

Societal Direction

Society should be (is) moving toward sustainability.

Economic activity will have to be planned.

The peace dividend was subverted.

The engineering labor force is changing with the entrance of more women and minorities.

A global economy is upon us.

Workshop Format

Papers were sent in advance so that participants could read them before coming to the workshop. Presentations by authors, therefore, were short in anticipation of focusing on major issues and questions. Small group discussions followed each paper presentation. Each group focused on one of the central questions that provided the integrative structure for the papers. Participants and facilitators rotated among the three groups so that participants had an opportunity to be in discussion with a variety of others in small group context. Most participants agreed that this format produced "much more conversation between papers than is usually the case at conferences and workshops."

Roles of Facilitators, Recorders, and Participants

The role of the facilitators was to keep the group focused and on track. The discussion began with participants in turn stating what they had learned or heard in the two preceding papers. This ensured that each person had the opportunity to participate. Recorders were used to capture the essence of each idea on a flip chart, creating a visual group memory. The visual record is important (1) to assure participants that their ideas have been heard, (2) to generate new ideas, (3) to focus the discussion on ideas, and (4) in cases of disagreement, to avoid personalizing conflicts.   The facilitators' role was to remain neutral, which required neither contributing nor evaluating ideas.   The role of the group members was (1) to keep the facilitator neutral, (2) to be good listeners, (3) to check for the accuracy of the group memory, (4) to contribute to the group memory, (5) to be sufficient and mindful of the time, and (6) to allow ideas to develop before mentally evaluating them.

Evaluation of Workshop

Evaluations of the workshop were conducted through four methods.   First, the representative of NSF and several participants wrote letters evaluating the strengths and weaknesses of the workshop.   Second, the interdisciplinary research team met after the workshop to assess the workshop and to plan for the revision and dissemination of papers.   Third, the facilitators provided input as to the process and how well it worked.   Fourth, a written evaluation survey was distributed to each participant at the end of the workshop.   All forms of evaluation revealed that the major objectives of the workshop were met to the satisfaction of the participants and interdisciplinary research team.   The quality of papers and presentations were rated high quality by 80% of the participants. That the process was successful is indicated by the high ratings of the participatory format and facilitators (86% and 78%, respectively).   Of the four integrative questions, participants felt we had best addressed gaps in our knowledge and how engineers will be impacted by economic conversion.   They felt there was less adequate resolution to how engineers will impact economic conversion and how engineers can be encouraged to participate constructively in it.   Not surprisingly given the diversity of participants and their backgrounds, the outcomes of the workshop received more variable ratings.   The outcomes of the workshop were of high quality to 79%.   One-fifth of respondents felt that development of the implications of economic conversion for policy, research, education, and professional engineering societies was less than adequate.

The research team and facilitators concluded that there was not enough time to move toward evaluation of ideas.   The discussion groups made varying degrees of progress in categorizing the ideas, but were unable to prioritize ideas nor to build consensus.   Much of the synthesis, therefore, was left to the interdisciplinary research team.   There was some tendency to defer to the authors of the papers as experts, which failed to acknowledge the expertise engineers gained through their lived experiences.   This pattern, however, may have been stylistic because many of the participants from academia were very verbal. We were surprised to find that there were disagreements over basic facts, such as how many engineers there are in the United States, and that these factual disagreements persisted throughout the workshop.   Many of these disagreements generated issues for the short-term research agenda.   As indicated in the pre-workshop questionnaire, there were significant differences in assumptions that could not be resolved in a workshop of short duration.

Appendix B
Selected Engineering Societies

American Association of Engineering Societies
415 Second Street, N.E.
Washington DC 20002
Ph. (202) 546-2237
Contact: Mitchell H. Bradley

American Chemical Society
1155 16th Street, N.W.
Washington DC 20036
Ph. (202) 872-4600
Contact: Dr. John K. Crum

American Consulting Engineer's Council
1015 15th Street, N.W., Suite 802
Washington DC 20005
Ph. (202) 347-7474
Contact: Howard M. Messner

American Institute of Chemical Engineers
345 East 47th Street
New York, NY 10017
Ph. (212) 705-7338
Contact: Dr. Richard E. Emmert

American Institute of Mining, Metallurgical and Petroleum
Engineers, Inc.
345 East 47th Street
New York, NY 10017
Ph. (212) 705-7695
Contact: Robert H. Marcrum

American Society for Engineering Education
11 Dupont Circle, Suite 200
Washington DC 20036
Ph. (202) 293-7080
Contact: F. Karl Willenbrock

American Society for Engineering Management
223 Engineering Management Building
University of Missouri at Rolla
Rolla, MO 65401
Ph. (314) 341-4560
Contact: Daniel L. Babcock, P.E.

American Society of Civil Engineers
345 East 47th Street
New York, NY 10017
Ph. (212) 705-7496
Contact: Edward O. Pfrang

The American Society of Mechanical Engineers
345 East 47th Street
New York, NY 10017
Ph. (212) 705-7722
Contact: David L. Belden

The Association of Energy Engineers
4025 Pleasantdale Rd, Suite 420
Atlanta, GA 30440
Ph. (404) 447-5083
Contact: Albert Thumann, P.E.

Association of Environmental Engineering Professors
c/o Dr. Bill Batchelor
Civil Engineering
Texas A&M University
College Station, TX 77843
Ph.(409) 845-1304
Contact: R. Noss

The Institute of Electrical and Electronics Engineers
345 East 47th Street
New York, NY 10017
Ph. (212) 705-7900
Contact: Eric Herz

Institute of Industrial Engineers
25 Technology Park/Atlanta
Norcross, GA 30092
Ph. (404) 449-0460
Contact: Gregory Balestrero

National Academy of Engineering
2101 Constitution Avenue, N.W.
Washington DC 20418
Ph. (202) 334-3200
Contact: William C. Salmon

National Society of Professional Engineers
1420 King Street
Alexandria, VA 22314
Ph. (703) 684-2800
Contact: Donald G. Weinert

The Society of American Military Engineers
607 Prince Street
P.O. Box 21289
Alexandria, VA 22320-2289
Ph. (703) 549-3800
Contact: B.G. Walter O. Bachus

Society of Women Engineers
345 East 47th Street
New York, NY 10017
Ph. (212) 705-7855
Contact: B.J. Harrod